PROFESSIONALISM AND POLICY: THE ROLE OF THE PROFESSIONS IN THE CENTRAL-LOCAL GOVERNMENT RELATIONSHIP

Professionalism and Policy: The Role of the Professions in the Central-Local Government Relationship

MARTIN LAFFIN

Gower

Published by
Gower Publishing Company Limited,
Gower House, Croft Road, Aldershot,
Hampshire, GU11 3HR, England.

and

Gower Publishing Company,
Old Post Road, Brookfield,
Vermont 05036, USA.

British Library Cataloguing in Publication Data

Laffin, Martin
Professionalism and policy: the role of the professions in the central-local government relationship.
1.Local government —— Great Britain —— State supervision
I.Title
354.4108'3 JS3111

Library of Congress Cataloging-in-Publication Data

Laffin, Martin, 1952-
Professionalism and policy.

1. Local government--Great Britain. 2. Social scientists in government--Great Britain. 3. Engineers in government--Great Britain. 4. Highway engineering--Great Britain--History. 5. Housing management--Great Britain--History. 6. Policy sciences. I. Title.
JS3137.L3 1986 352'.0004722'0941 85-21885

ISBN 0-566-00766-5

Printed in Great Britain by Paradigm Print, Gateshead, Tyne/Wear

Contents

Acknowledgements

This study forms part of a research programme established by the Social Science Research Council's Panel on Central-Local Government Relationships. As such, I am grateful to the Panel and to the SSRC for funding the research and to the members of the Panel for their advice and support during the research, in particular Mike Goldsmith and Ken Young. The research, too, would not have been possible without the help and co-operation of those local government officers and civil servants who talked, often with surprising frankness, about their roles in the policy process. Special thanks must go to those who were kind enough to read through and comment on the draft chapters.

The research was conducted while I was on the staff of the Tavistock Institute of Human Relations. I would like to thank my former colleagues for providing a hospitable environment, and especially John Friend to whom I owe a significant intellectual debt in this study. My thanks, too, must go to Ann Jamieson for typing and retyping drafts of this book with great cheerfulness, and to Jayne Moore, who typed the camera ready version.

But my greatest debt is to Julie Hadwin for her support and encouragement during the research and writing of this book and for relentlcssly applying her sub-editing skills to the text. It is to her that this book is dedicated.

Martin Laffin
Policy Studies Institute, London.

1 Introduction

Professionalism is one of the most significant features of British local government, while remaining one of the least researched. Conventional wisdom among writers on local government puts stress on the power and influence of the professional officers at local level despite the lack of systematic evidence. Considerable influence is also attributed to the professions in national policy making arenas but here again research has been very limited. The author has, already, investigated the nature of professional power within individual local authorities elsewhere[1], so it is an examination of the latter aspect of professional influence within the broader framework with which this study is concerned.

The aims of the research reported in this book are threefold. The first is to examine the conditions under which professionalism and individual professions have emerged within local government; in particular, it is argued that this process of professionalisation has to be seen within the context of the central-local government relationship. The second is to investigate the extent to which, and the conditions under which, professionals and professional groups have come to exercise influence within the policy process at the national level; and, although the concern of the study has been with the local government professions, the analysis should also contribute to an understanding of the policy influence of other professions outside local government. The third aim is to assess the significance for the relationship between central and local government of organised professional groupings that bridge the two levels of government.

The approach, then, to the study of policy in this book, focuses on those sources of policy initiation and influence which lie within government and its associated circles. This approach is in contrast to the more traditional approaches to the study of policy, which tend to concentrate on the legislative process and the role of pressure groups outside government in influencing policy. Broadly, this traditional approach explains policy outcomes in terms of the interplay among the political parties, interest groups and electoral support.

While no study can ignore these external pressures on government, the argument is that in many major areas of policy in the modern state, the main pressures for change or for policy maintenance are internal to government itself. These internal policy pressures have grown in importance as the modern state has increased in size, complexity and functional specialisation, and they take the form of groups of professional and other career officials, who act as distinct sources of policy initiation and influence within government.

So, in chapter two, the study begins by discussing two key concepts in the analysis of the policy process. The first is that of a policy community, which is defined as a relatively closed group of participants, including civil servants, professionals, politicians with a specialist interest in the policy area, and other interested actors. In analysing policy communities they are seen to have internal 'negotiated orders', and emphasis is placed on the significance of claims to knowledge, the value stances of participants and the relationships among them. Further, the environments of the communities will be seen as having three aspects - those public policy problems considered to fall within their ambit, relations with other policy communities and relations with outside constituencies, such as interest groups. The other concept is that of the issue network. These networks are looser collections of individuals who are outsiders rather than insiders and who share an interest in a given area of public policy.

Chapter three, then examines one particular and highly significant source of policy influence found within government - the 'organisational professions'. These 'organisational professions' are the product of the increasing range of services provided by the modern welfare state and the large organisations required to administer those services. These professions are considered as engaged in a continual process of negotiation over work tasks and claims to cognitive and normative competences. Their claims are articulated as accounts of professional practice which are devised to gain support from others and to assist practitioners in their work and give them a sense of shared concerns and interests. The chapter concludes with an explanation of the selection of two contrasting professions for further study and the research approach adopted.

The following two chapters look at the emergence of the two professions: highway engineering and housing management. In chapter four, the origins of municipal and highway engineering are traced back to the nineteenth century and the role of central government in the growth of the profession at this time can be seen. But the main story begins in the nineteen-thirties, when the basis for the later roads programme was laid. Against the background of the expanding roads programme, the growth of an influential professional community, embracing both central and local government, is described. However, by the early nineteen-seventies this community and the associated policy community faced new challenges from the developing anti-roads issue network and from cuts in public expenditure. For the local government engineers these challenges were accompanied by new difficulties as their professional claims became more difficult to sustain in their own authorities; and the responses of the engineers to these changes are discussed.

In chapter five, the origins of housing management are also traced back to the nineteenth century. But it was not until the nineteen-thirties that it began to emerge as a distinct occupational group within the public sector. The progress of the profession from this time is traced, particular attention being paid to the fluctuating interest of central government in housing management as opposed to house building. The major advances for the profession are seen as having taken place during the nineteen-seventies, after local government reorganisation and partly as a result of a central department's growing interest in housing management. During this period, the changing professional accounts are traced and recent initiatives in the training of housing staff are described.

The next four chapters detail the findings of four case studies dealing with issues that, in the light of the earlier chapters, are important ones for the two professions. Chapter six is concerned with county-district disputes over highway agencies, problems that arose after the 1972 local government reorganisation and the role that the engineers played in the national level attempts to resolve these problems and central government intervention in the setting up of two successive working parties. The other engineering case study, on highway maintenance, is presented in chapter seven. The developments in this field are followed back over a ten year period during which, as the roads programme declined, maintenance became of increasing interest to engineers. In particular, the chapter focuses on the reasons for the interest of local and central government engineers in road condition assessment systems and national road condition standards during this period. Chapter eight is the first of the housing case studies and looks at an unusual government committee, the Housing Services Advisory Committee, which contained a cross-section of people from the housing policy community. The reports published by this Group and its manner of working, are analysed. Chapter nine is concerned with the history of a piece of government legislation, the tenants' charter in the 1980 Housing Act. The origins of the tenants' charter are traced back to the emergence of the tenants movement and the new interest that both the political parties and the Department of the Environment civil servants began to take in tenant rights.

Chapter ten concludes the study by assessing its main findings in the light of the three research aims set at the start.

NOTES

[1] Martin Laffin, 'Professionalism and Party in Local Government: A Study of the Relationship between Professional Officers and Councillors in Eight Local Authorities in Northern England'. PhD Thesis, University of Bradford, 1983; Martin Laffin and Ken Young, 'The Changing Roles and Responsibilities of Local Authority Chief Officers', Public Administration, 63, 1, Spring 1985, pp.41-59; and Martin Laffin and Ken Young, Professionalism - Change and Challenge : The Local Authority Chief Officer in Modern British Politics, Allen and Unwin, London, forthcoming.

2 Analysing the policy process

POLICY COMMUNITIES AND ISSUE NETWORKS

Career officials within government have come to exercise considerable influence in the initiation, formulation and implementation of policy. Beer refers to this development as the emergence of the 'professional-bureaucratic complex'.[1] In the American context this complex is generally seen as composed of three insider groups - those politicians who have a special interest in the policy area, the bureaucrats from the relevant department and the interest groups representing the direct beneficiaries of the policy. These three elements have been referred to as a 'triple alliance' or as an 'iron triangle',[2] with the metallic metaphor stressing the closed nature of the policy process. These terms imply a relatively closed policy system or community to which non-members, even those who hold positions in other parts of the government machine, have difficulty in gaining access. Membership of the policy community is also small. While there are obvious differences between the US and the British case, the applicability of such a model to the latter is evident. It could even be seen as a more accurate description of the British than of the American system. Indeed, a growing number of commentators are arguing that an increasing number of decisions are being taken outside the formal parliamentary channels in Britain, which is now becoming a 'post parliamentary democracy'.[3]

This view of the policy process as dominated by closed policy communities has come under some criticism. Heclo, in particular, argues that while this view described American government in the 1950s and 1960s, it no longer represents an accurate description of the policy process there. He would like to replace policy communities with what he terms 'issue networks'.[4] Whereas policy communities imply the concentration of power in interest centred power structures, issue networks imply that power is much more dispersed. Thus, according to Heclo, the policy process has become more diffuse and the notion of issue networks better describes the situation where:

> ... for a host of policy initiatives undertaken in the last few years it is all but impossible to identify clearly who the dominant actors are. Who is controlling those actions that go to make up our national policy on abortions, or on income distribution, or consumer protection or energy? Looking for the few who are powerful, we tend to overlook the many whose webs of influence provoke and guide the exercise of power. These webs, or what I will call 'issue networks', are

> particularly relevant to the highly intricate and confusing welfare policies that have been undertaken in recent years.[5]

and:

> Increasingly it is through networks of people who regard each other as knowledgeable, or at least needing to be answered, that public policy issues tend to be refined, evidence debated, and alternative options worked out - though rarely in any controlled, well-organised way.[6]

Heclo sees no one as being in any sense in control of the emergence and formulation of policies in most policy areas, especially social policy. So power is not a quality that individuals possess but it is a quality of the 'configurations through which leading policy makers move and do business with each other'.[7] Unlike policy communities, issue networks are open and the barriers to entry are not difficult to scale: 'The price of buying into one or another issue network is watching, reading, talking about and trying to act on particular policy problems'.[8] Progress up, what could be called, the hierarchy of the network is dependent on the control that the individual has over necessary knowledge and, even more important, over the experts or those who are judged knowledgeable, the 'victuallers of knowledge in a world hungry for knowledge'.[9] But despite these shared assumptions on what constitutes knowledge, networks are not necessarily characterised by shared values or, in other words, a consensus on courses of action. In this, they are unlike policy communities to which the literature has generally attributed both a consensus on what is judged as knowledge and on objectives and action.

Although Heclo provides some valuable insights into the policy process, he does assume and exaggerate the apparent similarities among issue areas. Issues move around from being confined to narrow policy communities run by officials and experts, to partisan policy forums dominated by the ideological stances of the political parties and to policy debates falling within an extended network of interested parties.[10] The major aims of any policy making study must be to understand these differing and changing patterns, the reasons for particular patterns of policy making and their significance for policy outcomes. With these aims in mind, in the rest of this chapter the definition of the two concepts will be tightened up with a view to showing how the use of the concepts in conjunction can assist in the analysis of professional influence in government.

'Policy community' will be taken as referring to a relatively small group of participants which has formed to deal with some identifiable class of problems that have or could become the concern of central government.[11] This group invariably includes civil servants Jordan actually suggests that the concept should be defined as 'that comparatively small circle of participants that a civil servant might define as being of relevance of any particular policy'.[12] Other participants are likely to be politicians with a special interest in those problems, officials from government or semi-government agencies (especially the implementation agencies) and representatives from legitimised and 'respectable' interest groups. A major feature of communities is the limitation on access; not just anyone is invited or allowed to join. Those within the community generally operate quite stringent entry criteria, these vary among issue areas but include such criteria as possession of expert knowledge; occupancy of a senior position in a relevant organisation; what civil servants call 'soundness', meaning that the person can be trusted to observe the norms of the community; and reputation for 'getting things done'. Another important characteristic is a high degree of value consensus. This is because policy communities represent broad policy commitments for example, in the case of the roads community, to the continuation of the roads programme and, in the case of housing, to the involvement of central government in the improvement of the local housing system. Once that broad commitment is removed, the community usually disintegrates. So those serious conflicts that do occur within communities do not involve value issues fundamental to the community. Linked to the underlying value consensus, policy communities are characterised by relatively stable norms, rules and conventions governing the relations among the participants. Finally, the communities can have different purposes or functions at different times. At certain times in the natural history of an issue, the community may be primarily oriented towards the search for policy ideas, while at others it may be oriented towards action such as winning resources in competition with other policy communities holding different priorities.

In contrast, an 'issue network' refers to a loose collection of individuals who have a shared interest in an identifiable, but less well-defined, class of problems. They may or may not include officials or politicians, but the latter will constitute the major targets of network members' efforts to influence and persuade. Access to the networks is more open compared with the policy communities, it is more likely to be limited by the sheer number of people anxious to express their views than by exclusionary practices. But, as for policy communities, command over some resources, such as expertise, is important if the individual or group wishes to gain prominence and influence in the network. Consensus and stability are not prime characteristics of networks, they are much more likely to be affected by change than communities. For they encompass competing groups and, in particular, groups that often advocate change of a more radical type than would typically find acceptance in policy communities.

This account raises the possibility, indeed the likelihood, of policy communities and issue networks existing together in the same policy sectors. In fact, there are few cases where they do not co-exist. In a sense, then, a policy community can be seen as constituting the 'core' of an issue network. The policy community forms the focus of the network members' attention as they aspire to influence the presumed policy makers within the community and, moreover, the network members may themselves wish to join the community. However, there are variations among policy sectors in the relationships between the policy community and its equivalent issue network. Any analysis of these variations can start from the reasonable assumption that those within a policy community have a disincentive to widen the policy debate and, therefore, to bring others from an issue network into their policy discussions. This is suggested by Schattsneider's argument that those in positions of government power or influence seek to limit the participation of others, as it threatens to dilute their own power or influence, by maintaining low public visibility for their policy debates.[13] Visibility is important because the more those outside government know about the policies under consideration, the more likely they are to demand to participate themselves, when their own interests or concerns are involved.

This raises questions relating to those conditions under which, at least some, policy community members widen the policy debate and under which those in issue networks come to demand participation. Policy community members draw in others because they require further support or information, either because they wish to strengthen their sectional influence in the community or because they wish to increase the effectiveness of the community as a whole. These points will be illustrated in the analysis of the policy communities in the two policy sectors studied in this book. Meanwhile, those within issue networks characteristically attempt to enlarge the debate to include at least themselves by bringing pressure to bear on the relevant policy community. The form that this pressure takes varies widely. It may be limited to simply distributing literature to those interested, or it may take the form of attempts to mobilise wider public support through campaigns attracting the attention of the mass media. Again, in the course of this study, these different forms of pressure will be illustrated.

Although the most usual configuration is a policy community co-existing with an issue network, it is possible for one to exist without the other. A policy community may grow up in a highly technical area, in which only those immediately involved are interested. This may happen in the early days of a major technological innovation sponsored by government, outsiders may only become interested and begin forming networks when the wider, social implications of the innovation become known and public in some way. Nuclear power is a good example. In the early days of development, policy making could be restricted to the policy community and it is only more recently, when the potential dangers have become known, that an issue network has grown up. Another example will be detailed in this study, that of road building, which was similarly restricted to a small policy community until the late sixties, when an issue network of groups arose around the policy area. There is a marked, though not inevitable, tendency for such policy communities to emerge where a problem is recognised and defined almost solely at the level of central government and is seen as requiring an essentially 'technical' response or resolution.

The other main possible policy configuration is the growth of an issue network prior to the existence of a policy community. Typically, networks arise in response to the problem definitions of policy makers, either accepting them or attempting to influence them away from their original definitions. He who defines the issues, exercises power.[14] Those placed near or at the centre are likely to possess the most influence over the emergence and definition of public policy problems. However, some problems are recognised and defined outside the immediate circles of central government. In these cases, such problems are discussed and organised around and within issue networks; though it is possible, if the problem is not of widespread social concern, for it just to give rise to a single special interest group. For instance, the professional co-ordinating associations that have partly grown up in local government to act as networks to allow the exchange of ideas and experiences among individuals in similar social situations.[15] One recent example of such an emergent issue network is that concerned with the problems of correcting the effects of racial discrimination in local government recruitment.[16] This network is just beginning to coalesce around a definition of the problem that is now being recognised as an issue by those within government. Thus issue networks can emerge around what may initially be vague and ill-defined problems, their purpose being to clarify the issues and to mobilise action, either from the government or perhaps from other bodies. These networks too, especially in their early days, may be rent with disputes and even split up into more than one distinguishable network.

The members of an issue network may not even direct their attention towards government, but may be more concerned with their own immediate problems, so they may come to influence the policy making agenda without intentionally setting out to do so. For instance, it can be argued that child psychologists in the Freudian tradition, like John Bowlby, have had a tremendous impact on policy debates about the provision of child care facilities through their definition of the mother-child bonding process, although they have not organised themselves as a pressure group. This point highlights the value of thinking in terms of issue networks as well as of policy communities in explaining the emergence of public policies. For it stresses that the terms of a policy debate can be shaped by a wide range of influences, even though the final decisions may be taken by a small group of actors.

THE ENVIRONMENT OF A POLICY COMMUNITY

In the course of this section and the next, a framework for analysing policy communities will be outlined. Firstly, the environment of communities will be broken down into three types - constituencies, the contiguous policy communities and the operating environment.[17] Secondly, the internal organisation of the communities will be discussed and similarly resolved into three analytical categories - the normative, cognitive and the relational. The reason why the communities, but not the networks, will be analysed in greater depth is that the former are more central to the concerns of this study and because they are more discreet than the later. Of course, as will become apparent, the same categories can sometimes be helpfully applied to issue networks. Even so, most networks are too diffuse for them to be said to possess an environment in a meaningful sense.

The Constituencies or Sources of Responsibility

These are the set of public or other interests to which the actors in the policy community consider themselves to be responsible for the actions which they may select. The extent to which these other interests are able to influence or even control participants within a policy community varies. Some interests are unorganised and have no collective voice, others may be extremely well organised and capable of exerting considerable influence. At one extreme, participants in a community may take up these inarticulated interests and define their 'needs' for them, these needs definitions can reflect genuine social concern and, too, be used ideologically in bureaucratic battles. At the other extreme, a policy community can become dominated by an interest or interests outside government and, in this case, the interest usually manages to gain representation in the policy community. An often quoted example of this is the relationship between the Ministry of Agriculture and the National Farmers' Union.[18] There may be times when community participants attempt to distance themselves from overbearing or potentially overbearing constituencies, in particular from politicians trying to extend the arena of partisan debate into the community's policy area. The participants may respond by defining the area as 'technical' and challenging the right of the politicians to intervene.

It is quite likely that the participants at times face conflicting loyalties to their constituencies. Civil servants may find themselves torn between loyalty to their department and their sense of obligation to a client group. Local government officers may similarly find themselves having to choose between their allegiance to a profession and to an employing authority. Though, at the same time, it is quite possible that participants acquire a sense of allegiance to the community itself which may tend to displace their original allegiance or allegiances. So, clearly, an analysis of the pattern of allegiances must be an important guide to uncovering those external influences on participants.

Contiguous Policy Communities

Insofar as certain policy issues are perceived as extending across the boundaries of communities, participants engage in communication or in transactions across these boundaries. Membership of these contiguous policy communities is not mutually exclusive and participants sometimes play roles in more than one community. Moreover, attempts to resolve problems by participants in one community often have important consequences for other contiguous communities. For example, as will be seen later in this study, decisions taken on economic policy have had considerable impact upon housing and roads policy. So, in attempting to explain events in a community, note should be taken of the intentions and actions within contiguous policy communities.

The Operating Environment

This refers to those aspects of the environment of a policy community from which problems emerge that can be classified as falling within its recognised action space.[19] Action space connotes that class of problems which is generally regarded as the business of the policy community. It bears the same relationship to policy community as domain does to an organisation.[20] The operating environment includes both social and material elements. For instance, in formulating future road building plans, participants have to take account not just of the problems of engineering design but also the changing social demand as expressed in changing patterns of car and vehicle usage. Though it should be stressed that participants do not necessarily perceive the operating environment in the same way. Thus a research study should take into account these possible perceptual differences and consider their implications for the policy stances of participants.

THE INTERNAL ORDER OF THE POLICY COMMUNITY

While the stability of policy communities has been stressed, in this section this stability will be seen as resting on shared understandings that emerge among participants. The internal stability of communities is, for the most part, not maintained through formal, constitutional conventions but through informal and often unspoken conventions and understandings that have grown up over time. Participants become socialised into policy communities and their cultures - the informal rules governing the communities. As the two policy communities under study will illustrate, it is typical for membership of a policy community to change slowly and for new members to be introduced gradually into it so they have the opportunity to absorb the shared understandings that have accumulated within the community - a process of socialisation.

Essentially, policy communities are based on more or less tacit agreements to confine conflict in the interests of joint action. This is not necessarily to say that all participants enjoy equal influence in determining policy outcomes. Clearly they have largely to accept existing power relations as given in their negotiations with other participants in the community. As Strauss and his collaborators show in their formative study of the negotiated order at the organisational level of a hospital:

> ... there is a patterned variability of negotiation in the hospital pertaining to who contracts with whom, about what, as well as when these agreements are made. Influencing this variability are hierarchical position and ideological commitments.[21]

In this study of a psychiatric hospital, Strauss and his colleagues emphasise the fragility of the social order and the 'necessity for continually reconstituting the bases of concerted action or social order'.[22] On the larger scale of policy communities similar efforts have to be expended to ensure their continuation. These efforts are pointed up by Heclo and Wildavsky in their study of the financial community within Whitehall:

> There is no escaping the tension between policy and community, between adopting actions and maintaining relationships, between decision and cohesion, between governing now and preserving the possibility of governing later. To cope with the world without destroying the understandings their common life requires - this is the underlying dilemma facing the community of political administrators.[23]

The rest of this section will be concerned with an analysis of the internal workings of policy communities and the problems involved in their maintenance. As a first step, three aspects of their internal order will be identified - the cognitive, the normative and the relational. These three categories are intended to be of analytical value, in particular in highlighting the role of claims to knowledge in the policy process. Clearly, in practice, it can be difficult to make these distinctions, that between the cognitive and the normative raises especially difficult problems on occasion. However, the analytical value of the categories lies in the way in which they direct attention towards the different aspects of the process whereby problems in public policy come to be recognised as such within communities, then defined and acted upon.

The Cognitive Order

This refers to what passes for accepted knowledge within a policy community, the distribution of that knowledge and who manages the experts within it - who are the 'experts in the use of experts?'.[24] Policy communities are characterised by agreement on what constitutes relevant and accepted knowledge and also, what might be termed, cognitive conservatism towards the acceptance of new knowledge or information. Those within a policy community have to make judgements about the relevance and epistemological acceptability of new knowledge and, by the same token, the purveyors of new knowledge have to persuade community participants that any new knowledge satisfies a community's criteria of relevance and epistemological acceptability. The first criterion, that of relevance, is fairly self-explanatory; but it should be noted that it usually changes over time as policy makers' problem perceptions change.

The second criterion, epistemological acceptability, requires more explanation. Accepted knowledge can either be demonstrated or accredited. New knowledge, or what is potentially new knowledge, can be assessed on its merits as it stands, in other words its status as knowledge has to be demonstrated, but this involves time, effort and often a degree of technical understanding. These cognitive limitations on policy makers create a tendency for them to rely upon external sources of accreditation in determining the validity of potentially new knowledge. Policy makers have come to rely heavily on these 'accredited sources of knowledge' which they have come to accept as authoritative. These sources include the implementation agencies, researchers from the universities, members of recognised professions and those pressure groups that have come to be accepted as respectable.

The professions are a prime example of an accredited source. The possession of qualifications recognised and sanctioned by the institutions of higher education and other socially prestigious bodies gives credibility to a profession's claims to expertise. Some professionals are so well accredited that policy participants are willing to accept most of the individual professionals' views without inquiring further into the premises behind those judgements, that is without asking for them to be demonstrated as correct. However, it should also be noted that professional claims above all represent claims to the possession of knowledge rather than just information; in other words, a profession claims to have a systematic body of knowledge enabling its members to understand certain types of policy problem.

In this context, claims to knowledge can be seen as being based on different types of accredited source. Professionalism, then, represents one type of source but there are others. Another way of saying this is that there are different and often competing types of rationale, which participants or potential participants in the policy process draw on to justify their claims to knowledge. Thus, one popular rationale is for a participant or group of participants to argue that they possess knowledge not shared by others because of their close involvement in the 'frontline' of service delivery. For example, local government officers may use this argument in policy discussions with civil servants when the officers' professional claims are disputed or not seen as relevant by the latter. The problem with this type of claim is that it is vulnerable to the counter-claim that knowledge of specific authorities is not the same as knowledge of authorities in general. A further possible objection is that such knowledge is unverifiable and that it is quite likely that someone else will arrive at different conclusions from their own first hand experience. Given the inherent weakness of claims based on this type of rationale, the acceptability of these claims depends more than is the case with professional claims, on the claimant's personal standing in the eyes of others within the policy community.

The Normative Order

This refers to the interplay of values within a community and to those shared values that underlie the continued existence of a community. There are two general aspects to the normative order - the differences between the values found inside and those found outside the community, and the particular values that underpin the normative consensus within it. The first aspect refers to the criteria of entry to a community, while the second relates to the negotiations over values and the compromises among divergent values that arise within it. To work together, participants do not necessarily have to share the same 'basic values' but only similar 'intermediate values'. The notion of 'intermediate values' is introduced to emphasise the possibility of community participants coming to agree with each other over the desirability of a particular policy goal from differing, or even incompatible value stances.

In this study 'values' are defined as 'standards of the desirable which men apply in making choices'.[25] It should be stressed that that which is just desirable is not necessarily the right course of action. For a course of action to be right, it must conform to a wider moral principle. Neither is the right course of action a reflection of the interests of those involved. In the case of many policy participants, it makes little sense to say that they are acting in their own interest when their own personal interests are not at stake. This is the case with many pressure group activists, for example, who campaign on behalf of deprived groups to which they themselves do not belong. Similarly, civil servants occasionally take up the interests of such groups, as they define them, without apparently being motivated by career considerations.

Basic values rarely, if ever, originate within policy communities. The purpose of policy communities is not the formation of value positions but the resolution of or compromise between competing values. In other words, communities are characterised by agreement on intermediate values but not necessarily on basic values. Of course, basic values are formed in the course of individuals' upbringing and education and, during their membership of employing organisations, occupational groups, political parties and so on. Values formed in these settings are carried into communities and one approach to understanding the value stances of participants is to refer back to their social origins and their organisational and occupational affiliations. These affiliations are usually of particular significance as they represent constituencies which may be placing pressure on them.

Turning to the interplay among values within a community, as a general rule, participants seek to make their own values dominant in the policy process. Those who occupy strategically placed positions in the relational order are clearly in a better position than those participants less favourably placed; these others will have to rely solely on persuasion to achieve their ends. However, if these relatively disadvantaged participants begin to strengthen their position, they risk exclusion on the grounds of breaking the 'rules of the game'. This touches on the important distinction between procedural values (this is how things ought to be done) and substantive values (these are the ends that ought to be pursued). Agreement on procedural values - which can be vague, general and assumed rather than articulated - can sustain order despite profound differences over substantive values. So agreement on procedural values is conducive to agreement on substantive values. This refers back to Heclo and Wildavsky's point about how participants, in particular civil servants, face a tension between policy and community. Indeed, some actors may see their role simply in procedural terms as maintaining the community regardless. Dominant participants can refer to apparent abuses of procedural values in justifying the exclusion of groups or individuals with unacceptable substantive values from participation in policy discussions.

The Relational Order

This refers to the pattern of relationships within a policy community pertaining to who negotiates with whom, about what and when. Given differences in political and organisational positions among participants, they have varying access to, what might be termed, power resources. The civil servants involved, for example, usually act as gatekeepers in controlling a community's dealings with the other policy communities whose co-operation is essential for the achievement of some of the aims contemplated in the first community. Clearly this places them in a powerful position as few, if any, other participants enjoy the same access to other parts of the governmental machine. This is certainly the case for local government officers, though less so for councillors who have party political channels available through which to bring pressure to bear on other policy communities.

There are, too, other power resources available to different participants. One important resource is the power of those who are responsible for implementing policies or decisions. Other participants quite often have to listen and take seriously the views of implementors as, unless the latter's co-operation or even just indifference is ensured, a policy could be effectively undermined by them at the implementation stage. Local government officers, for example, within a policy community can use this type of threat of non-co-operation by themselves and their colleagues.

Under the heading of relational order, some note should be made of Heclo's strictures, quoted earlier, concerning the value of searching for dominant actors. These strictures were based on his study of the formation of social policy in the US, Sweden and Britain, where he failed to detect the presence of particular policy makers who could be said to dominate the process. The findings of this study support a similar conclusion in the case of the two public policy fields investigated. It became apparent that different participants exercised different types of power or influence at different times. Nevertheless, power within these two policy sectors was not as diffuse as Heclo's approach implies, but tended to centre on the policy communities.

POLICY COMMUNITIES: OVERVIEW

Although this analysis of the internal order of policy communities has dealt with each aspect separately, clearly these aspects are closely inter-related. The possession of accepted knowledge can endow the holder with influence that may substitute for the holder's lack of organisational or political power relative to other actors. Equally, those with weaker claims to knowledge may be able to compensate through the exercise of organisational or political power; for example, through their control over key access points to other policy communities. Meanwhile, the underlying relationships within a community have to be explained by reference to the values held by participants. Against this background, power resources and claims to knowledge can be seen as instrumental resources used by actors in attaining those ends they value.

Similarly, the other major analytical distinction, between a policy community and its environment, should not be drawn too sharply. Again, the nature of the boundary between community and environment tends to shift over time and differ between policy sectors and, in particular, it may be more or less permeable. The permeability of communities depends crucially on the changing climate of opinion in informed circles, which usually includes the appropriate issue networks, and also on ideological shifts in the wider society. Thus, although the formal policy makers or decision takers may inhabit the policy community, the agenda for the policy debate may be shaped outside its boundaries. As Schon argues:

> Underlying every public debate and every formal conflict over policy there is a barely visible process through which issues come to awareness and ideas about them become powerful. The hidden process by which ideas come into good currency gives us the illusory sense of knowing what we must worry about and do.
>
> We pay attention to the visible conflicts over policy; but by the time such conflict has crystallised, issues have long since been identified, ideas for solution have long since been available, sides have been defined and taken. These antecedent processes are as crucial to the formation of policy as the processes of science are crucial to the formation of plausible hypotheses.[26]

This raises the possibility that there are some circumstances where such an overwhelming consensus can arise that power is effectively drained away from the policy community concerned; in other words, the terms of the policy debate become so well defined by those outside that policy conclusions become almost inevitable. Some commentators appear to assume that this is nearly always the case, that the policy agendas of communities are inherently responsive to changes in public opinion.[27] But there is no evidence for such inherent responsiveness. The process of problem identification and definition and, then, policy response, is typically a political and competitive process. As Edwards and Batley observe in their study of the development of the British urban programme, the process is one of 'interaction and conflict between the perceptions and interests of the groups involved', in their case, 'the inhabitants of deprived areas, voluntary groups both of the interest and promotional kind, other community activists, politicians, political parties, government departments, academics, the government, etc'.[28] In the two policy sectors investigated in this study, the process is characterised by competing problem definitions and policy proposals, though there were important differences in response between the two policy communities. In housing the policy community formed to take the lead in initiating and promoting change in local authority housing policies; while those within the roads policy community resisted new and emerging policy definitions that they saw as threatening their existing policy commitments.

POLICY COMMUNITIES AND THE CENTRAL-LOCAL GOVERNMENT RELATIONSHIP

Local authorities form centres of political power in their own right, with their own electoral bases. With the expansion of the welfare state in Britain, they have come to fulfill a major role as the local deliverers of services. Indeed, the very factors that have given rise to the proliferation of policy communities, particularly the increasing size and specialisation of government have also involved a growth in the number and range of activities or functions carried out by local authorities. But, given their nature as 'miniature political systems',[29] they have a pronounced tendency to develop divergent approaches to the implementation of those services they provide. This tendency creates a potential tension within the relationship between central and local government. Central government is inevitably concerned with local government services for both financial and political reasons. Financially, central government is the major contributor to local government spending. Politically, there is a growing tendency for interest groups and the public to hold central government responsible for locally provided services and a widely held view - among national and local politicians, civil servants and professional officers - that services should be provided on a uniform and territorially just basis. This means that there are powerful pressures on central government to intervene in local government service provision. Thus local authorities are subject to **constraints** on their freedom of action and are the **targets** for new policy initiatives from the centre. At the same time, they are themselves a **source** of policy influence and initiative for those services provided by local government.

Policy communities play a key role in this process. Actors within both central and local government with similar policy concerns often have a shared interest in the establishment and maintenance of policy communities. They are typically based on functional rather than territorial divisions so they cut across the central-local government divide. Thus they tend to counteract the centrifugal tendencies of localism by bringing actors from central and local government closer together. For those at the centre, policy communities are a means both of gathering information and ideas and of co-opting key local actors. Clearly, they require information and advice from those in local government to ensure that their perceptions of the problems and any policy responses are appropriate. As important, they often need the assistance of professional officers and others, in order to formulate an adequate account of the problems and for ideas about how to go about solving them. Moreover, they need to know what the likely reaction of local authorities would be to different types of central initiative and to be able, at the end of the day, to present any new policies as emerging from joint central-local deliberations in order to encourage the co-operation of authorities.

Meanwhile, for those in local authorities, policy communities have three main functions. Firstly, they provide channels through which those from local government can persuade their central counterparts not to make policies that are locally unworkable or likely to create further serious problems. In this way, local and central actors have a common interest in the orderly regulation of central-local relations. This is expressed in the conventions that have grown up in this field; for example, the local authority associations expect to be consulted first on all except the most minor policy changes. Secondly, communities are sometimes seen as a means of mobilising central support for the attainment of local objectives, particularly by local government officers concerned to gain acceptance by their elected members of more 'enlightened' policies. Thirdly, participation in policy communities is a major way of influencing policy and of achieving professional or, indeed, party political goals, for it should not be forgotten that elected politicians are sometimes important members of a community. For those in local government, both officers and councillors, the possibility of such influence has made such participation very worthwhile.

In conclusion, it should be stressed that there are important variations in policy making patterns. In particular, policies relating to the services delivered by local government do not always emerge from policy communities. Thus some policy sectors come to be dominated at particular times by the political parties with their ideological commitments, so that the role of the policy community becomes marginalised. The language of political debate within the policy sector changes too, from being predominantly technical to being ideological. Occasionally, a major political change, usually a change of the party in government, involves such a major shift away from the broad policy commitments underpinning the existence of the community, that it is disbanded by the new policy makers or disbands itself. This has been the case with the housing community investigated in this study, which failed to outlast the outgoing Labour Government. Other, more gradual shifts in political and social attitudes can also affect policy communities. Such shifts typically involve changes in how problems are defined, changes which policy communities have to respond to, either by trying to incorporate them into their own account of what the problems are, or by resisting through reasserting their own views. In the main part of this study, the case of the roads community will illustrate the ways in which a policy community goes about the defence of its own policy commitments.

NOTES

[1] Samuel H Beer, 'The Adoption of General Revenue Sharing : A Case Study in Public Sector Politics', Public Policy, 24,2, 1976, p.160.

[2] Beer, 'The Adoption of General Revenue Sharing', p.160; and Hugo Heclo, 'Issue Networks and the Executive Establishment', in A King (ed), The New American Political System, American Enterprise Institute, Washington DC, pp.87-124.

[3] For example, J J Richardson and A G Jordan, Governing Under Pressure : The Policy Process in a Post-Parliamentary Democracy, Martin Robertson, Oxford, 1979.

[4] Heclo, 'Issue Networks'.

[5] Ibid, p.102.

[6] Ibid, p.104.

[7] Ibid, p.88.

[8] Ibid, p.102.

[9] Ibid, p.103

[10] Cf A Grant Jordan, 'Iron Triangles, Woolly Corporatism and Elastic Nets : Images of the Policy Process', Journal of Public Policy, 1, 1981, p.108.

[11] This concept is based on the similar but not quite identical one of 'policy system' in J K Friend, J M Power and C J L Yewlett, Public Planning : The Inter-Corporate Dimensions, Tavistock Publications, London, 1974, Ch.2.

[12] Jordan, 'Iron Triangles', p.105.

[13] E E Schattsneider, The Semi-Sovereign People, Dryden Press, Hinsdale, Illinois, 1960, especially Ch.1.

[14] Cf Schattsneider, The Semi-Sovereign People; Robert Paul Wolff, 'Power', in his The Poverty of Liberalism, Beacon Press, Boston, 1968; and Peter M Hall, 'A Symbolic Interactionist Analysis of Politics', Sociological Inquiry, 42, 1972.

[15] See chapter three.

[16] This example was suggested to the author by Ken Young.

[17] Following Friend et al, 'Public Planning', pp.27-30.

[18] For an analysis of this relationship see Graham K Wilson, Special Interests and Policy Making : Agricultural Policies and Politics in Britain and the United States of America, 1956-70, John Wiley, London, 1977.

[19] Cf Friend et al, Public Planning, p.29.

[20] As used by James D Thompson, Organisations in Action, McGraw-Hill, New York, 1967, pp.27-29.

[21] A Strauss, L Schatzman, D Erhlich, Bucker R and M Subshim, 'The Hospital and the Negotiated Order', in E Friedson (ed) The Hospital in Modern Society, Collier-MacMillan, London, 1963. For an elaboration of this approach, see Anselm Strauss, Negotiations : Value Contexts, Processes and Social Order, Jossey-Bass, London, 1978.

[22] Cf Strauss et al, 'The Hospital and the Negotiated Order', p.165; and Hall, 'A Symbolic Interactionist Analysis of Politics'.

[23] Hugh Heclo and Aron Wildavsky, The Private Government of Public Money, MacMillan, London, 1974, p.xv.

[24] Heclo, 'Issue Networks', p.103.

[25] Philip E Jacob and Hames J Flink, 'Values and Their Function in Decision Making', supplement to American Behavioral Scientist, 5, 1962, p.13.

[26] Donald A Schon, Beyond the Stable State, Temple Smith, London, 1971, p.123. See also his 'Framing and Reframing the Problems of Cities' in David Morley, Stuart Proudfoot and Thomas Burns (eds), Making Cities Work, Croom Helm, London, 1980, pp.31-65.

[27] For example, R N Cobb and C Elder, Participation in American Politics : The Dynamics of Agenda Building, John Hopkins University Press, Baltimore, 1975; and William Solesbury, 'The Environmental Agenda', Public Administration, 54, 1976, pp.379-399.

[28] The Politics of Positive Discrimination : An Evaluation of the Urban Programme, 1967-1977, Tavistock, London, 1978, p.220.

[29] Jeffrey Stanyer, Understanding Local Government, Fontana, London, 1976.

3 Professionalism in government

INTRODUCTION

While the last chapter outlined a framework for the analysis of policy sectors, this chapter will analyse the bases of professional power within these sectors. This analysis will encompass both the underlying trends that have led to the professionalisation of policy sectors and to the grounds of professional power within professionalised sectors. The chapter, then, will draw on the extensive sociological literature and use the insights contained in some of the literature, to develop a theoretical perspective on the growth of professionalism in government and professional influence in the policy process.

PROFESSIONS AND PROFESSIONALISATION

The conventional approach to the sociology of the professions centres on the question of which occupations are or are not professions. A great deal of effort has been expended in the search for an agreed definition of 'profession' but with little success. Typically, these efforts have taken the form of setting out the traits or attributes which, once acquired, qualify an occupation for professional status. Usually these sets of attributes are based on one of the more successful professions that emerged during the nineteenth century, more often than not medicine is taken to be the ideal-typical profession. The attributes listed usually include some or all of the following: the possession of an established body of systematic knowledge; a commitment to altruistic service to the client; the existence of an occupational association, the membership of which is the licence to practise; a high level of educational achievement among practitioners; and considerable autonomy at work.[1]

This approach has been widely criticised, not least because it assumes a consensual and static view of society. The clear implication is that if only an occupation can attain a given set of attributes it will automatically be able to assume the mantle of a 'profession'. But the process whereby an occupation becomes a profession is considerably more complex and politically contested than this approach implies. In this process, the relationships between a profession and other social and political power centres are crucial, rather than just the profession's internal characteristics. Recent studies of the professions, including the two contained within this book, demonstrate the central importance of these relationships in the professionalisation process.[2] Indeed, the formation of internal attributes in large measure reflects a profession's strategies in negotiating support and deflecting challenges from other power groupings. Professional attributes, then, tend to vary from profession to profession and according to changing social and political circumstances. Thus the weakness of the conventional approach is that any defining list of attributes, as Johnson observes, tends 'to be derived from the analysis of a very few professional bodies and include features of professional organisation and practice that find full expression only in Anglo-American culture at a particular time in the historical development of these professions'.[3] Similarly, Freidson argues that the problems of defining 'profession' have arisen through attempts to 'treat profession as if it were a generic concept rather than a changing historical concept, with particular roots in an industrial nation strongly influenced by Anglo-American institutions'.[4]

It is more fruitful, then, to follow Freidson and Johnson in emphasising the relationships between a profession and other power centres within society, rather than its supposed attributes. Accordingly, professionalism should be defined as a 'peculiar type of occupational control rather than an expression of the inherent nature of particular occupations'.[5] Thus what is distinctive about a profession compared with other occupations is that it has been given the 'right to control its own work'. [6]

The central and defining aim of a profession is the maximisation of its autonomy or freedom from control by others, both within the immediate work setting and in the institutionalised regulation of the relations between the professionals and the consumers of their services. This aim is justified in terms of the possession of special competences on the part of the professional practitioners. The other aims of the profession - such as enhanced incomes, social status and control over other occupations - are subsidiary in that they are expected to follow on from increased autonomy. Meanwhile, what are often thought of as the defining attributes of professionalism, such as long periods of training and qualifications, can be interpreted as strategies used in the process of winning autonomy.

This approach to professionalism directs attention towards the question of 'what are the circumstances under which people in an occupation attempt to turn it into a profession, and themselves into professional people'.[7] In other words, the interesting questions about professionalism are those relating to the process of professionalisation. This process can be defined as one whereby 'an organised occupation, usually but not always by making a claim to special esoteric competence and to concern for the quality of its work and its benefit to society, obtains the exclusive right to perform a particular type of work, control access to it and control the right of determining and evaluating the way the work is performed'.[8] So professional status is typically pursued through the establishment of claims to certain exclusive cognitive and normative competences, claims which have to be accepted by other powerful political and social interests. In consequence, these interests' perceptions of professional claims constitute the social reality within which a profession is located. In particular, those occupations that work within or predominantly within government, typically professionalise by negotiating claims to competence in the development and implementation of policy within the appropriate policy community.

THE ORGANISATIONAL PROFESSIONS

In this context, it is important to note that professionalism is increasingly an organisational phenomenon. Professions are now found predominantly within large scale organisations and especially within governmental organisations. The professional ideal, underlying most of the early attempts at definition in terms of attributes, of professionals as independent practitioners selling their services to clients within a professionally controlled market has become an anachronism. The working conditions of all the professions have changed as large governmental and commercial organisations have come to dominate modern society. Even the older and well established professions, like medicine, have come to be practised within organisations. Though it should not be forgotten that even in some of the older professions, and engineering is a good example [9], a large number of practitioners have always been salaried employees working in organisations. However, the newer professions - like planning, social work and teaching - are even more distinctively organisational in their origins and current working conditions. This has led Larson to argue that all modern professions, and even the older professions, are to an increasing extent 'organisational professions':

> All professions depend to a certain extent on large organisations and on the state - if nothing else, because all have a teaching arm. All professions are, today, bureaucratised to a greater or lesser extent. Organisational professions should not be seen, therefore, as sharply distinct from older and more independent professions, but as clearer manifestations of tendencies also contained within them.[10]

The concept of an organisational profession may seem paradoxical as professionalism and bureaucracy are often considered as fundamentally incompatible. It is often argued that professional decisions are too complex and difficult to be controlled through the characteristically bureaucratic control mechanisms of rules and hierarchical authority. So these types of decision require a degree of work autonomy incompatible with bureaucracy.[11] However, the evidence indicates that bureaucracy and professionalism should rather be seen as complementary forms of social control - professionalism tending to emerge where bureaucratic controls are proving ineffective.[12]

The conditions necessary for the rise of professionalism within organisations lie in the emergence of substantial areas of discretion within them and the consequent problems of organisational control. Areas of discretion are created in organisations where there is no scope or only limited scope for the routinisation and, therefore, bureaucratic control of an organisation's work. The potential for routinisation of this work is limited by increasing technological complexity and the problems of supervision inherent in large scale organisations.[13] Individual employees in many areas of organisational activity have to be permitted to make choices among possible courses of action or inaction where they frequently encounter exceptional situations not easily accommodated within any set of organisational rules.[14] The need to control or regulate the resulting exercises of discretion has given rise to the professionalisation of work within organisations; 'professionalism, in other words, makes the exercise of discretion predictable'.[15]

Professionalism has become a distinctive feature of the modern welfare state because many of the tasks involved in service provision are not susceptible to routinised control.[16] Differences in the types of task required in the modern service state have given rise to two major forms of professionalism - the public service and the technobureaucratic.[17] The first type, the public service, has arisen as a direct result of the expansion of state intervention, which has brought many new tasks into government that were previously unrecognised or performed by private citizens. These professions are those like teaching, social work, housing management and planning, which are involved in welfare or regulation in the 'public interest'. It is these professions that are in the front line of service provision and usually have a client orientation of some kind.[18] Their claims to professional status are means of legitimating their claim to autonomy or freedom of control by others, generally their bureaucratic superiors and their clients.

The second category, the technobureaucratic profession, is generated by the necessity of managing and administering the large scale organisations in which the public service professions are employed. These professions do not have a client orientation but are oriented towards those who head their organisations. Essentially, their claims to professional status are used to legitimate their exercise of hierarchical authority over subordinates, including public service professionals and their influence over policy in relation to those in a position of, at least formal, power over them, in particular the elected politicians. While some professions are purely technobureaucratic, like public sector accountancy, other professions are a mixture of the public service and the technobureaucrat. The trend has been for the public service professions, as their work organisations have grown in size, to develop individual career paths that involve movement into technobureaucratic positions. Indeed, there has often been resistance from the public service professionals to be subordinated to technobureaucratic professionals who do not share the same public service professional background as themselves. For example, the British Association of Social Workers has objected to the appointment of those not qualified in social work to posts of social services director; similarly, but more successfully, ex-teachers have managed to maintain a monopoly of education directorships.

THE BASIS OF PROFESSIONAL POWER

Professions, then, have emerged within government as centres of power with some degree of autonomy, in some cases quite considerable, from direct political and bureaucratic control. They have autonomy in three distinct senses: in the immediate work situation, in the ability to be self-regulating or self-governing at the level of the profession and in being autonomous sources of influence on the formation and implementation of government policy. Clearly these different types of autonomy are closely associated. Generally the process of professionalisation involves increases in professional autonomy at all three levels. As such, professionalisation is a political process of obtaining and enlarging on the acceptance by others of a professions' claims to exclusive competences or expertise in certain areas of social concern. This section and the next will consider this process and emphasise the function of professional accounts within it, these accounts being analysed in terms of two types of claim to exclusive competences - the cognitive and the normative.

An account is a collection of statements which are made by a social actor or group of actors, to explain or justify a course of action, or a desired course of action.[19] Accounts are typically used either where there is a discrepancy between the aspirations of actors and the reality of their situations as they perceive them, or where there are serious challenges to their status or position in society. More specifically, a professional account is a statement and justification of the particular competences which a profession claims and of the set of tasks which its members regard as falling within their jurisdiction. A professionalising occupation has to arrive at an account of its practice which is sufficiently convincing to win both the consent of its members and of those in a position to support its aspirations. Those professionals in the forefront of the formation and modification of professional accounts often face difficult problems in trying to reconcile the expectations of those within and those without a profession.[20] These leading professionals are defined in this book as the 'professional community' as distinct from the profession as a whole.[21] It is members of this community who are engaged in the process of negotiating with the representatives of other social and political interests in order to win support for and reflect challenges to their professional claims.

The leaders of a profession, then, typically try to manage the emergence of problem perceptions and definitions within the public policy arena to lead other actors to conceive of the services of the profession as essential in handling certain problems of social significance. As Rueschemeyer observes, 'the professions do have, in varying degrees, a special voice in determining normatively what constitutes a problem fit for, and in need of, expert intervention; thus doctors influence our conceptions of illness which separates that special state from a condition of normal social responsibility; and lawyers shape our views of justice and order, fostering, for instance, a more formal and procedural than substantive conception of these values'.[22] They tend to modify or reinforce their account of professional practice in accordance with their perceptions of the supports offered and the challenges posed by other actors. More specifically, they have to demonstrate both the relevance of and the accredited status of their claims to knowledge and also their normative competence or reliability in carrying out the tasks involved. Clearly the ability to win support from strategically placed actors is crucial.[23] These actors include senior civil servants, government ministers, the members of other, more established, professions and leading councillors from the local authority associations.

Obviously, one important group of actors is that found within the appropriate policy community. Their acceptance of a profession's case is clearly essential. Certainly, no one outside the community is going to accept that case if those already regarded as the 'experts' within government have rejected it. However, their acceptance is not sufficient in itself as policy communities do not control certain of the resources necessary for professionalisation. A professionalisation project is also dependent on the acceptance by those controlling these other resources of the case for the profession in question. So that bodies - like universities and other educational institutions, the local authority associations, the Privy Council (for royal charter status), the Department of Education and Science and individual employing organisations who may have to fund training initiatives - have to be convinced of an aspiring profession's case.

Undoubtedly these 'insider' actors play a crucial role in the establishment and maintenance of professional status and influence. However, the growth of 'outsider' groups representing social and political interests, such as tenants' organisations, amenity groups and consumer groups, has significant implications for the professions. They are increasingly finding themselves in competition with these groups which have their own accounts of policy problems and solutions. Often these groups form issue networks, as those who are searching for and beginning to assert an alternative definition of the problems in a policy sector feel their way towards each other. Established professions try to cope with these challenges by adopting strategies such as co-option or exclusion; examples of these differing strategies will be considered in the case of the road engineers in chapter four.

So far, the professions have been treated as if they displayed a remarkable degree of internal homogeneity. However, the members of a profession are seldom all agreed on an account of professional practice. So that, as Bucher and Strauss observe, 'the assumption of relative homogeneity within the profession is not entirely useful; there are many identities, many values and many interests'.[24] Characteristically, professions are divided into segments or sub-groups reflecting sub-specialisms, different work locations, employing organisations, locations in organisational hierarchies and so on.[25] A major division is that between the public service and technobureaucratic branches of a profession. The nature of the different work tasks and different hierarchical locations in organisations create a tendency for the evolution of differing professional accounts. The public service professionals tend to develop professional accounts oriented towards the client rather than the employing organisation, towards enabling the professionals to cope with street level situations and gaining countervailing power against the employing organisation.[26] In contrast, technobureaucratic professionals tend to form accounts aimed at coping with co-ordinative and management tasks, and at legitimating their exercise of organisational power over their subordinates and influence with their political masters. In addition, the technobureaucratic segment of a profession, which also includes the professional community, also tries to sustain an account of the professional practices of the public service segment; this is both to maintain their control over subordinates and to use in policy negotiations with policy makers.

This is not the only significant type of division that occurs within professions in the government sector. The members of a profession also tend to identify themselves with their employing organisation, out of a sense of loyalty as well as self-interest. So that conflicts between employing organisations often create tensions and divisions within the profession whose members they employ. As will be seen in the main part of this study, this has been the case on occasion between the engineers employed by central government and those employed by local government. Some mention, too, should be made of those professions whose membership bridges the private and public sectors. Again, this builds in potential tensions and possible conflicts of interest. In particular it has been argued that professions which bridge the two sectors act as channels of influence for the private sector over the public sector.[27]

PROFESSIONAL CLAIMS TO COGNITIVE AND NORMATIVE COMPETENCES

In this section, professional accounts will be analysed in terms of types of claim to competence - the cognitive and the normative. This analysis is relevant in understanding not only the professionalisation process whereby an occupation comes to be granted powers of self-regulation, but also the degree of autonomous influence that a profession is able to exercise in the policy making process.

Claims to Cognitive Competence

This refers to professional claims to possess and be able to apply knowledge that is relevant and accepted and also exclusive. The conventional accounts of professionalism have usually stressed the key importance of the possession of a body of systematic knowledge in the assertion of professional claims. However, this type of account misses the real cognitive strength of professionalism. That strength lies precisely in that knowledge which cannot be systematised, for the systematisation of knowledge means expressing it in the form of rules or applicable principles. Once in this form, knowledge is much more easily transmittable so that cognitive exclusiveness becomes more difficult to maintain. Crozier makes this point well in arguing that a profession's success is self-defeating:

> ... the expert's success is constantly self-defeating. The rationalisation process gives him power, but the end results of rationalisation curtail this power. As soon as a field is well covered, as soon as the first intuitions and innovations can be translated into rules and programmes, the expert's power disappears.[28]

Jamous and Peloille's account of professionalism encapsulates this tension well. They define a profession in cognitive terms simply as an occupation 'whose indetermination/technicality ratio, intrinsic to the systems of productions is generally high'.[29] Technicality refers to that type of knowledge that can be understood and communicated to others in the form of rules or, less precisely, that can be articulated without too much difficulty; that is those aspects of professional knowledge that the conventional approach stresses. Indeterminacy refers to that type of knowledge that cannot be systematised or codified in the form of rules. It is this indeterminate or tacit knowledge that, according to Wilensky:

> ... helps explain (the professions') achievement of exclusive jurisdiction; it also helps explain their traditionalism. The client public sees a mystery which is not given the ordinary man to acquire. Such tacit knowledge is relatively inaccessible,

it is also less subject to direct criticism and quick change.[30]

Thus the greater the proportion of a profession's knowledge base that is in the form of tacit or indeterminate knowledge, the more likely it is to be able to maintain its cognitive exclusivity.

However, a profession's indetermination/technicality ratio does not remain constant over time. Like other social institutions, the professions, if they are to survive, have to respond to changing pressures from within their environments. There are general social changes like the rising level of education which mean that ordinary citizens are more able to scrutinise critically the cognitive presumptions of the professions. Technological developments also change the indetermination/technicality ratio; one good example is the impact of computerisation on accountancy.[31] Professions, too, in competition with each other often question one another's cognitive claims when they are involved in jurisdictional disagreements. One example is the resistance that town planning faced from the engineering profession to its emergence as a distinct profession, the engineers challenged town planning for its lack of a 'scientific' knowledge base.[32]

The pressures for codification come as much from within as without professions, though ultimately even the internal pressures derive from the necessity of providing a standardised, and therefore marketable, service to the consumer - whether employer or client. Codification serves a vital function in standardising services as it sets up 'a transcendent cognitive and normative framework within which, ideally, differences in the interpretation of practice and in the definition of the "commodity" can be reconciled'.[33] Such a framework is necessary for the communication of professional knowledge to new entrants to a profession and to structure the debate within a profession on the nature of the services they are performing and, in particular, to lay down boundaries between what is and what is not regarded as acceptable professional practice. Without such a framework to contain the fissiparous tendencies within a profession, it would be unable to sustain a distinctive identity. It would mean that consumers would hesitate to engage the services of practitioners because they would be uncertain about what type of service was offered - whether it was relevant to their problems and likely to be effective in solving them. Thus some codification is an essential precondition to the accreditation of professional knowledge, the other major precondition being the qualification sanctioned by the state in some way. Under these conditions, consumers generally accept professional claims to knowledge as accredited and do not expect those claims to be demonstrated each time before they use professional services.

There is, too, an important compensatory strategy that often goes hand in hand with codification, that of mystification. While codification lowers the barriers between the professionals and lay people, mystification raises them. It means increasing the apparent indeterminacy surrounding professional work, for instance by obscuring the reasoning behind a decision or piece of advice by making it unnecessarily complicated or by expressing it in technical jargon. However, this strategy can prove counterproductive if it simply alienates people, for a profession is dependent on a basic minimum level of co-operation from lay people if it is to retain its status and influence. In other words, it must continue to have a basis in whatever is the prevalent 'commonsense'.[34]

Claims to Normative Competence

The other type of competence to which professions lay claim is that of being able to discern and observe the right actions in the course of their work. They must be not only knowledgeable but also trustworthy. It is this very knowledgeability, in the form of the possession of exclusive cognitive competence, that creates the need for some guarantees of trustworthiness on the part of the professions. For the recipients of professional services, whether clients in the more conventional sense or government policy makers, are themselves not adequately knowledgeable to solve the problem or to assess the service received. However, as has already ben mentioned, the problems of social control posed by the modern professions arise not only from their possession of expertise but also from the problems of social control inherent in large and complex organisations. Thus where external controls are believed to be impracticable and undesirable, professions adopt 'a code of ethics or some other publicly waved banner of good intentions'[35] to assure others of their trustworthiness and ethical integrity. As with cognitive assurances, the acceptability of such ethical assurances depends a great deal on the social assumptions about professionalism.

The nature of these normative competences or ethical codes to which professions lay claim is analysable in terms analogous to Jamous and Peloille's indetermination/technicality ratio. Thus some of the principles in these 'codes' are relatively determinate in that they specify precisely what actions or behaviours are unprofessional; for example, breaking client or employer confidentiality or accepting sexual favours from clients or patients. The enforcement of these rules or principles can be monitored by those outside a profession. However, insofar as a profession can claim to be observing an 'ethical code' in the sense of good intentions that cannot be codified into such a determinate form, such monitoring from without the profession becomes more difficult. In fact, the assumed need for the profession to make difficult ethical judgements in the immediate work setting becomes an argument for professional autonomy at the levels of both professional self-regulation and in the work setting.

A profession has to find a balance between codifying its good intentions and leaving them implicit. A minimum amount of codification is necessary if a profession is to convince others that its members are trustworthy and that it is committed to internal discipline. The extent of codification tends to vary with the variations in the moral credibility of the profession in the eyes of others; codification tending to increase with the perceived need to reassure others that the profession remains trustworthy. In addition, and for similar reasons, there are internal pressures within professions for normative as well as cognitive codification. A basic framework of rules provides at least an initial guide to new entrants into a profession and gives existing practitioners, especially when they are involved in ambiguous and anxiety inducing situations, some guidance and a structure within which to discuss the ethical dilemmas in professional practice.[36] Moreover, as was mentioned in the last section, a normative framework is as important as, if not more so than, a cognitive framework in standardising the professional 'commodity' and so assuring consumers of the homogeneity of a profession's service.

Although professions recognise that there are ethical dimensions to their work, characteristically they tend to understate the extent to which their members are involved in making normative judgements. They present their judgements as 'technical' in nature and resist attempts by others to define these judgements as value choices. The extent to which professions can maintain 'technical' or 'scientific' definitions of their work depends on the type of work involved and how this work is perceived by others. Generally, if the work is seen as essentially relating to the physical or material world, then presenting it in non-normative and 'scientific' terms is relatively unproblematical. However, when professional work is seen by others as relating to the social or human world, defining it as other than normative is difficult. These differences in perceptions reflect generally held assumptions about the world. Whereas these assumptions admit the possibility of scientific laws governing the physical world, they do not readily admit the same possibility in the case of the social world. So those professions whose tasks are seen as lying predominantly in the social world are typically seen by others as involving evaluative not empirical judgements. Accordingly, it is difficult for these professions to establish claims to cognitive competences in areas of work which are not regarded as requiring the application of such competences.

Finally, professions also have to maintain a basic level of 'value congruity' between their internal values and those of the wider society.[37] This is sometimes a precarious process when there are major shifts in public opinion or when scientific progress is so rapid that it outstrips the public debate on the moral implications of new discoveries.

PROFESSIONALISM IN THE CENTRAL-LOCAL GOVERNMENT RELATIONSHIP

The main part of this book is concerned with the role of professional communities in the context of the central-local government relationship. A professional community, as has been noted, is understood here as comprising not the profession as a whole but the elite or leadership of the profession. It is these professional officers who are participating or seeking to participate in national level policy discussions and who are prominent in sustaining and modifying the dominant account of professional practice. The national policy making arenas to which they belong or aspire to belong are usually the policy communities. But there are significant variations among the different policy communities in the relations between professional and policy communities. Within some policy sectors a professional community can come to form a major part of a policy community as it comes to dominate a sector; this will be seen to have been the case in the roads sector. In yet other sectors professional communities can be almost totally excluded from the policy community where it has come to be dominated by other powerful interests; this will be seen to have happened in the case of the housing policy sector.

The other main policy concept, that of an issue network, is also relevant in understanding how professions emerge within local government. Occupations can be viewed as professionalising through the initial crystallisation of an issue network around a set of issues of social concern. Typically, networks take the form of a co-ordinating association which 'primarily exists for specialists to discuss operational and occupational problems within a narrow context, but it has no sanctions or authority mechanism and it does not seek to bestow a qualification on its members through an examination process'.[38] It is only later on that such an association acquires, if at all, other functions such as the qualifying, protective and educative;[39] many of the local government professions illustrate this pattern of development, engineering and accountancy are good examples.[40] Co-ordinating associations arise then, where officers are employed by different authorities but come to recognise the shared nature of their problems and work experience. It should be stressed that this initial recognition, and any later assumption of a professional mantle, is hastened by the fact that professionalism is a dominant cultural norm in local government, as indeed it is in the rest of British society.[41] Certainly, any new occupational group within local government has to assert itself within the context of this norm and so, almost inevitably, is obliged to adopt at least some of the forms of professionalism.[42]

Moreover, central government policy towards local government has a strong built-in bias towards the professionalisation of services. Again, this represents a strong cultural preference,[43] reflecting partly the view that local government falls within the realm of the 'specialist' involved in the execution of policy not the making of policy,[44] and partly the tendency of the national political and official elites to see professionalism as a countervailing force to the forces of localism.[45] The two policy sectors studied, roads and housing, both provide good illustrations of the importance that these elites attach to professionalism. Professionalisation, then, has placed limits on the control of local authorities over their staff, especially as qualification, training and initial selection have been entrusted to outside agencies such as the universities, training colleges, professional institutions and central training boards. This has taken place with the encouragement of central policy makers, who favour these outside sources of professional accreditation over locally based training and experience.

So the emergence of professional autonomy in terms of self-regulation has to be understood within the context of the central-local government relationship. Equally, the emergence of professions as autonomous sources of influence within the policy process has also to be understood within this relationship. Professional communities tend to evolve their own policy objectives which may not necessarily coincide with those of their employers, though in any case they usually have a significant influence on their employing authorities' policy objectives. Typically, they pursue these objectives within both their home authorities and through national level channels - the professional associations, the local authority associations, contacts with pressure groups and informal contacts with civil servants and other policy makers. In particular, they use these channels to influence the distribution of decisions between central and local government. On some issues they attempt to persuade central policy makers to give local authorities discretionary powers in order to increase their members' local discretion; while on the other issues they try to persuade the policy makers to increase the constraints on local authorities in the belief that this will strengthen the position of their colleagues locally. Examples of attempts by professionals to manage the distribution of decisions between local and central government in both these ways will be found in this study.

The rest of this book will examine the process of professionalisation and the exercise of policy influence by professions through a detailed analysis of two contrasting professions within local government. The research approach adopted was exploratory and the two professions were selected on the basis of dissimilarity in order to illustrate the wide range of factors relating to the development and influence of the professions in government. The two professions selected for study are highway engineering and housing management. The following outlines indicate their main characteristics and points of contrast:

> **Highway Engineering** possesses an apparently hard 'technical' and 'scientific' cognitive base and, ostensibly, there appear to be few normative issues raised by engineering work. Organisationally it is well entrenched within local authorities, especially at the county level, and also inter-organisationally entrenched as its membership encompasses central and local government to a greater extent than the other professions. Finally, it has had to respond to serious challenges from amenity interests and environmental pressure groups over recent years.
>
> **Housing Management** lacks a hard 'technical' cognitive base and does face serious difficulties in convincing others that it has special competences; its existing occupational tasks tending to be seen by lay people, such as local councillors, as simply the application of commonsense. Organisationally and inter-organisationally it is much less well entrenched than engineering. But, like highway engineering, it has been challenged by new interest groups in the form of tenant and other groups. Finally, the choice of housing management has meant that a contemporary, albeit constrained, process of professionalisation could be studied and also contrasted with an earlier example of professionalisation in engineering.

The rest of this study consists of two chapters which analyse the historical development of the two professions in the light of the theoretical considerations contained in this chapter; and four further chapters which form more detailed case studies of the policy role of the professions. Chapter six, on highway agencies, considers the problems faced by a profession when its membership is divided by organisational loyalties and is faced by an issue that is not easily defined as 'technical' but has been created by the existence of competing territorial jurisdictions. In contrast, chapter seven looks at the issue of highway maintenance, focussing on how the profession responded to pressures from without by codifying professional practice in a field where professional knowledge previously had been tacit in nature. The next two chapters on housing reveal a policy community in which the professionals had come to play a considerably more marginal part than was the case in the roads policy community. Chapter eight describes the workings of the Housing Services Advisory Group, on which the professionals were poorly represented and in which the politicians and civil servants generally took the initiative. Similarly, in chapter nine, the professionals are seen to have had little influence in the emergence of the tenants' charter, a major new initiative in the field of housing management; instead the elected politicians were the main initiators.

NOTES

[1] Cf Wilbert E Moore, The Profession : Roles and Rules, Russell Sage, New York, 1970, pp.6-22.

[2] See, in particular, Terence Johnson, Professions and Power, Macmillan, London, 1972; and Magali Sarfatti Larson, The Rise of of Professionalism, University of California Press, Berkeley, 1977.

[3] Johnson, Professions and Power, p.26.

[4] Eliot Freidson, 'The Theory of Professions : State of the Art', in Robert Dingwall and Philip Lewis (eds), The Sociology of the Professions : Lawyers, Doctors and Others, MacMillan, London, 1983, p.22.

[5] Johnson, Professions and Power, p.45.

[6] Eliot Freidson, Profession of Medicine, Dodd Mead & Co, New York, 1970, p.71.

[7] Everett C Hughes, The Sociological Eye, Aldine Atherton, Chicago, 1971, p.340.

[8] Eliot Freidson, 'Professions and the Occupational Principle', in Eliot Freidson (ed), Professions and their Prospects, Sage, Beverley Hills, 1974, p.22.

[9] See Edwin T Layton's history of engineering and its relationships with employing organisations, The Revolt of the Engineers, Case Western University Press, Cleveland, 1971.

[10] Larson, The Rise of Professionalism, p.179.

[11] For the classic statement of the conflict see W Richard Scott, 'Professionals in Bureaucracies', in Howard M Vollmer and Donald M Mills (eds), Professionalisation', Prentice Hall, Englewood Cliffs NJ, 1966, pp.265-75.

[12] Quite a large number of studies provide empirical confirmation of this, see in particular: A Hastings and C R Hinings, 'Role Relations and Value Adaptation : A Study of the Professional Accountant in Industry', Sociology, 4, 1970, pp.353-66; George A Miller, 'Professionals in Bureaucracy : Alienation among Industrial Scientists and Engineers', American Sociological Review, 32, 1968, pp.755-68; and Celia Davies, 'Professionals in Bureaucracies : the Conflict Thesis Revisited', in Dingwall and Lewis, The Sociology of the Professions, pp.177-94.

[13] Cf 'When the overall responsibility of the organisation cannot be broken down into fairly routine specialised tasks ... expert judgements of professionals rather than disciplined compliance with the commands of superiors must govern operations in the interests of efficiency.' Peter M Blau and W Richard Scott,

Formal Organisations : A Comparative Approach, Routledge and Kegan Paul, London, 1963, p.247.

[14] For a more developed version of this argument, see Charles Perrow, 'A Framework for the Comparative Analysis of Organisations', American Sociological Review, 32, 1967, pp.194-5.

[15] Larson, The Rise of Professionalism, p.198.

[16] Though the professionalisation of services may be more marked in the Anglo-American context. Professionalism appears to be a more pronounced characteristic of British society compared with Europe. For example, in the case of manufacturing industry, excessive professionalisation has impaired the efficiency of British compared with German industry, according to J Child, M Jones, I Glover and P Lawrence, 'A Price to Pay? Professionalism and Work Organisation in Britain and West Germany', Sociology, 17, 1983, pp.63-78. There is clearly scope for parallel research in the case of social policy.

[17] Larson, The Rise of Professionalism, p.179.

[18] For the most part these are the 'street level bureaucrats' indentified by Michael Lipsky in his Street Level Bureaucracy, Sage, Beverley Hills, 1980.

[19] See Marvin B Scott and Stanford M Lyman, 'Accounts', American Sociological Review, 33, 1968, pp.46-62.

[20] Michael D King discusses such tensions in 'Science and the Professional Dilemma', in Julius Gould (ed), Penguin Social Science Survey 1968, Penguin, Harmondsworth, 1968, pp.34-73.

[21] This is a departure from the conventional view of a profession as a whole being a 'community' within a 'community', the best statement of this approach is W J Goode, 'Community within a Community : The Professions', American Sociological Review, 22, 1957, pp.194-200.

[22] Dietrich Rueschemeyer, 'Professional Autonomy and the Social Control of Expertise', in Dingwall and Lewis, The Sociology of the Professions, p.53.

[23] Cf Eliot Freidson, 'A profession attains and maintains its position by virtue of the protection and patronage of some elite segment of society which has been persuaded that there is some special value in its work. Its position is thus secured by the political and economic influence of the elite which sponsors it.' Profession of Medicine, p.72.

[24] Rue Bucher and Anselm Strauss, 'Professions in Progress', Journal of Sociology, 66, 1961, pp.325-34.

[25] Ibid p.325; and Rue Bucher and John G Stelling, Becoming Professional, Sage, Beverley Hills, 1977, Ch.1.

[26] See Lipsky, Street Level Bureaucracy.

[27] Patrick Dunleavy, Urban Political Analysis, MacMillan, London, 1980, pp.113-7.

[28] Michael Crozier, The Bureaucratic Phenomenon, University of Chicago Press, Chicago, 1964, p.165.

[29] H Jamous and B Peloille, 'Changes in the French University Hospital System', in J A Jackson (ed), Professions and Professionalisation, Cambridge University Press, Cambridge, 1970, p.112.

[30] H Wilensky, 'The Professionalisation of Everyone?', American Journal of Sociology, 70, 1964, pp.148-9. Cf Donald Schon's distinction between 'knowing-in-action' (tacit knowledge) and 'knowledge-in-action' (determinate or technical knowledge) in The Reflective Practitioner : How Professionals Think in Action, Temple Smith, London, 1983, pp.21-75.

[31] P D Montagna, 'The Public Accounting Profession : Organisation, Ideology and Social Power', in Eliot Freidson (ed), The Professions and their Prospects, pp.135-51.

[32] G E Cherry, The Evolution of British Town Planning, Leonard Hill, London, 1974, pp.165-69. For a useful discussion of inter-professional rivalry in terms of cognitive claims, see W Goode, 'Encroachment, Sociology and Medicine', American Sociological Review 25, 1960, pp.902-14.

[33] Larson, The Rise of Professionalism, p.40.

[34] Cf Burkart Holzner, Reality Construction in Society, Schenkman Books, Cambridge Mass, 1968, p.128.

[35] Freidson, Profession of Medicine, p.134.

[36] The importance of professionalism in serving practitioner's psychic needs has generally been overlooked by sociologists despite its apparent significance, see I E P Menzies, The Functioning of Social Systems as a Defence Against Anxiety, Tavistock Institute, London, 1970.

[37] The term 'value congruency' is taken from Holzner, Reality Construction, p.134. L J Sharpe makes a similar point in relation to the need for the local government professions to keep their 'operating ideologies' in line with public opinion, 'Instrumental Participation and Urban Government', in J A G Griffith (ed), From Policy to Administration, Allen and Unwin, London, 1976.

[38] G Harries-Jenkins, 'Professionals in Organisations' in J A Jackson, Professions and Professionalisation, p.60.

[39] Ibid, p.60.

[40] See T L Poynton, The Institute of Municipal Treasurers and Accountants : A Short History, IMTA, London, 1960.

[41] Cf Child et al, 'A Price to Pay?'.

[42] Though the more recent signs, as will be seen in the course of this study, are that there is growing scepticism of and resistance to new professional claims in local government.

[43] A cultural preference which has been manifest in other policy areas. In particular, Michael Burrage argues that the professional ideal, widely shared by politicians from both parties and by civil servants, has influenced the way in which the nationalised industries in this country have come to be controlled. In particular, they are given a tremendous amount of autonomy in practice 'compared with comparable nationalised industries in other countries'. 'Nationalisation and the Professional Ideal', Sociology, 7, 1973, pp.253-72.

[44] Cf J F Warren, 'The tasks with which modern local government began were largely of an executive character, and beyond all doubt the primary need of the local authorities in recruiting officers they could place in executive charge was for professional men. At the very outset, it should be noted, there was a contrast between the needs of the local authority and the needs of central government, since the task of the latter did not extend to very much in the way of direct provision, but rather to the functions of control and regulation, and was thus to be conceived of as administrative in character rather than as executive.' The English Local Government System, Allen and Unwin, London, 1963, p.193.

[45] In Kaufman's terms, professionalism counteracts the centrifugal tendencies of localism. Herbert Kaufman, The Forest Ranger, John Hopkins University Press, Baltimore, 1960, Ch.7.

4 The development of highway engineering

INTRODUCTION

This chapter is concerned with the development of one of the oldest professions in local government, engineering, the origins of which lie in the nineteenth century. The chapter will begin by considering the reasons why the profession emerged and, in particular, the parts played by the different interests at both national and local government levels. After this initial discussion, the focus of the rest of the chapter will be on one of the main engineering specialisms, that of highway engineering. This focus reflects the fact that, since the mid nineteen fifties, highway engineering has been the most important field of local and central government engineering activity in terms of both finance and central-local relations. The growth of a professional community, encompassing local and central government, will be charted against the background of the expanding road building programme. In the later part of the chapter the different challenges to this professional community will be discussed. And finally, the responses of the policy and professional communities to these challenges will be examined.

THE NINETEENTH CENTURY ROOTS OF MUNICIPAL ENGINEERING

The nineteenth century origins of municipal engineering are inextricably tied up with the emergence of two major social problems - that of the insanitary living conditions of the urban working classes and the poor condition of the country's road communications. The former was undoubtedly the more important in the nineteenth century and ever since municipal engineering has been closely associated with public health in many people's minds. The public health problems of that time arose directly out of the rapid industrialisation of England and the resulting concentration of industry and population in major urban centres. The overcrowded living conditions, the absence of waste disposal systems and of a pure water supply all contributed to the spread of disease. Smallpox, cholera, typhus and tuberculosis were endemic in this country. But by the middle of the nineteenth century, some doctors and others began to relate the spread of these diseases to impure water supplies, poorly built and ventilated accommodation and to the accumulation of refuse and waste on the streets. Notably these problems were seen by the reformers as aspects of the more fundamental problem of public health. Further, this problem was regarded by them as requiring, by the standards of the time, a massive increase in the scope of government intervention because they considered that much of the disease and illness was preventable. Once it becomes known that a social problem can be solved or ameliorated, the social and political pressures for government intervention increase.

The full extent of the public health problem and the possibility of remedying it reached official attention in the 1842 Report on the Sanitary Condition of the Labouring Population.[1]

The report was authored by Edwin Chadwick, a leading reformer and, at that time, the Secretary to the Poor Law Commission. Among other recommendations, the Report stressed the importance of appointing properly qualified officers and of efficient administration in the improvement of sanitary conditions in the cities. In the Report, Chadwick and many of his witnesses strongly criticised the existing public commissions and local boards for being amateurish, dominated by local vested interests and lacking co-ordination among the different services:

> Witnesses of the most extensive practical experience lay the greatest stress on the necessity of lifting these important branches of administration out of the influence of petty and sinister interest, and of doing so by securing the appointment of officers of superior scientific attainments, who (subject to a proper local as well as general control) may be made responsible for directing any new expenditure on a scale of efficiency as well as of economy. A competent, scientific, and efficient management, let it be applied to what part of these works it may, can scarcely fail to be immediately as well as ultimately the most economical management.[2]

Chadwick concluded that 'securities should be taken that all new local public works are devised and conducted by responsible officers qualified by the possession of the science and skill of civil engineers'.[3]

This influential Report led to the passing of the first Public Health Act in 1848. The Act provided for a General Board of Health and gave powers to local health boards which, in incorporated towns, were the existing town councils. The Act also established the post of sanitary engineer and surveyor under the protection of the Board of Health whose permission was required before any borough could dismiss their engineer.[4] However, this protection was lost when the Board was abolished in 1858, and 'the protection from undue local influence and intimidation the Surveyor had enjoyed for 10 years'[5] was lost.

This loss of statutory support from a central agency forms an important part of the background to the formation of an occupational association of municipal engineers. But it was not until 1871 that municipal engineers formed an association for the first time. The discussions leading up to the Municipal Corporations Act 1872 provided the stimulus and the opportunity for the municipal engineers to organise themselves to press for the return of the protection that they had previously enjoyed.[6] However, they were unsuccessful in persuading Parliament. Nevertheless, this experience suggested that a permanent occupational association could have an important role to play, so, in 1873, the Association of Municipal and Sanitary Engineers was founded. In the terms mentioned in chapter 3, this was initially a co-ordinating association not a qualifying association. At this time too, as befits a co-ordinating association, it included only senior officers - 'civil engineers and Surveyors holding permanent appointments under the various urban and rural sanitary authorities within the control of the Local Government Board'[7]; assistant surveyors were not included until 1886. It is clear from the objects of this association that the early municipal engineers saw the problems of public health as their area of activity.[8] The first major change in the new association did not take place until 1885 when it acquired a qualifying function. It was the first of the engineering associations to institute an examination.

The second major social problem was the poor state of road communications. In the early years of the nineteenth century, the condition and administration of the country's roads was chaotic. Even major routes were poorly maintained. There was no central agency responsible for maintaining a national network of roads and capable of enforcing the repair and maintenance of major highways. The administrative framework was essentially that laid down in the Tudor period and still depended on the voluntary help of local notables and the statutory labour of local people. The appointment of county surveyors was in the hands of the Justices of the Peace in Quarter Sessions, the administrative forerunner of the modern county councils, under the Statute of Bridges Act 1530. But the quality of these surveyors varied widely and, in doing their duties, they relied heavily on the willingness of local notables to take the initiative, which was a serious weakness.

Notably, the impetus for change came from two quarters outside the rudimentary local governments - the military engineers and certain of the private turnpike trusts. The political necessity of subduing and anglicising Scotland and Ireland had led to major road building projects in these countries at the end of the eighteenth century and the early nineteenth.[9] Thomas Telford (1757-1834), one of the great civil engineers and a founder member of the Institution of Civil Engineers,[10] acquired his reputation as a road-builder in constructing roads in the Scottish Highlands, where he became the Commissioner for Highland Roads. On his return to England, be became the first <u>civil</u> engineer to be appointed by Parliament, when he became the Commissioner for the Holyhead-London Road.[11] It appears that the condition of this road had given rise to such strong feelings in Parliament, among those members who had to use it frequently, that the then customary aversion to central government intervention was overcome. Though it was a further five years before Parliament voted the necessary funds for the road.

Telford's next proposed civil engineering project was less successful. He proposed to straighten out the Great North Road between London and York, in order to improve travelling times and the comfort of travellers. But many of the towns along the existing route, hearing that they might be side-tracked by a new road, petitioned Parliament, which subsequently rejected the proposal. At the same time, the railways were beginning to be constructed on a large scale, linking up more and more urban centres. This effectively killed off any idea of major road building. It was only after the invention and application of the internal combustion engine that roads came back into their own.

As well as introducing purely engineering improvements, Telford was also concerned with the calibre and training of engineers responsible for road maintenance and construction.[12] This was a concern shared by the other great reforming road engineer of the time, John Loudon Macadam (1756-1836). Unlike Telford, Macadam was not originally a military engineer, nor was he associated with any major construction projects. His interest in the field arose from his concern with the state of the roads in his neighbourhood, which led him to become involved in several turnpike trusts. Macadam promoted two types of improvement through these trusts, technical improvements to the actual riding surfaces of roads, and administrative improvements to the organisation of road maintenance and construction. He argued that roads should be designed to carry vehicles rather than vehicle weights limited to what the existing roads could bear, and he pointed out that travelling speed and comfort depended on the state of the road surface. Accordingly, he further argued that the weight-bearing capabilities and travelling qualities of roads could be improved considerably through the construction and maintenance of a 'strong, smooth, solid surface' on the roads.[13]

The second type of improvement advocated by Macadam is of more interest to the historian of local government manpower policies. He considered the problem of poor roads as not simply resulting from technical ignorance of what constituted an adequate road surface, but also from relying on voluntary, unpaid and untrained labour both in the administrative and labouring work:

> Gratuitous services are ever temporary and local; they are dependent on the residence and life of the party; and have always disappointed expectation. Skill and executive labour must be adequately paid for if expected to be constantly and usefully exerted; and, if so exerted, the price is no consideration when compared to the advantages to the public.[14]

Accordingly, Macadam:

> ... persistently urged that the essential feature of any efficient administrative service must necessarily be the trained professional official, paid a salary which enabled him to live without corruption - Macadam said he must be of good social status - giving his whole time to his duties, and held responsible by the governing body for the success of the undertaking.[15]

This must be one of the first examples of the advocacy of a paid, professional service at local level. It, of course, preceded Chadwick's similar views by several years. Both Macadam and Chadwick saw the professionalisation of locally-based services as offering some guarantee of the commitment of officials to the task in hand, and to their efficiency and impartiality when under pressure from local vested interests. Views that the Webbs applauded as 'scientific' in their history of the King's Highway.[16]

Macadam appears to have influenced new legislation. A year before his death, the Highways Act 1835 was passed by Parliament. This Act made provision for each parish council to become an autonomous highway authority with the power to elect and pay a surveyor; similarly, in the same year, the Municipal Corporations Act, which created many new corporations in the towns, also give the corporations highway responsibilities. But Macadam's influence was perhaps greatest through the diffusion of good practice through his work in turnpike trusts. This work became widely known among other trusts and local councils where the officials began to use 'macadamised' road surfaces.

Although within the towns the engineering profession became well-established with a wide range of public health and highway responsibilities, outside in the rural areas local administration remained primitive. Highway responsibilities were divided between the county surveyors, appointed by the Quarter Sessions, and the many small local authorities. This division of powers was not clearcut, and it left many major roads in the care of small and inefficient local authorities. So, by the 1880s there was growing Parliamentary interest in the reform of local government in the countryside. This interest sparked off the formation of the County Surveyors Society in 1885, in conditions remarkably similar to those that had led the municipal engineers to organise themselves fifteen years previously.[17] Finally, administrative reform came in the Local Government Act 1888, which created the first county councils. The new counties were obliged to maintain those roads which they had adopted as 'main' roads from the existing local councils and those roads that were disturnpiked, leaving those councils to look after the remaining 'minor' roads. The creation of the new county councils then, meant that the surveyors had their range of responsibilities enlarged and they began to emerge as an independent force within local government. Three years after the formation of the county councils, the earlier municipal engineers' association recognised their county equivalents and the association changed its name accordingly to 'The Institution of Municipal and County Engineers'.[18] The County Surveyors Society, however, maintained its separate existence, which has continued up to the present day.

During the latter half of the nineteenth century civil engineers enjoyed considerable prominence within their localities. Certainly in the towns, much of the improvement in public health was the result of the activities of the municipal engineers in building new systems for the water supply, sewerage disposal, rubbish disposal and in improving the condition of the urban roads. Similarly, in the rural areas, the county surveyors became third in importance to the county clerks and the treasurers in the newly created county councils.[19]

ROADS AND THE INTERNAL COMBUSTION ENGINE

Towards the close of the nineteenth and the beginning of the twentieth century, there was a renewal of interest in the state of the nation's roads. The increasing numbers of bicycles and motor vehicles created a new public with a strong interest in improving the conditions of the roads. The arrival of the motor vehicle, in particular, meant that travellers became more conscious of road surface conditions, while the road surfaces became subject to heavier wear. The official reaction had been simply to impose weight and speed limits on the new vehicles, partly in the face of the vocal anti-car lobby and partly to limit expenditure. The first pro-roads lobby was formed not by the owners of motor vehicles but by cyclists, though the association was later taken over by motorists. This new association, the Road Improvement Association, was formed in 1886.[20] The aim of the association was not the construction of major new roads but the improvement of the road surface and of road alignments and gradients. Given this aim, one of the main activities of the Association was the promotion of the use of tar in surfacing roads and in 1907 they arranged for demonstrations of tar spreading machinery.[21]

The problem of poor road conditions was not seen simply as a technical problem but also as an administrative problem. Road conditions varied widely among the counties and central government, in the form of the Local Government Board, had failed to intervene to impose any national standards or to co-ordinate highway policies across the country.[22] Thus:

> No effective national leadership or advice was available in connection with the considerable changes needed in the techniques of road construction, although it must have been obvious that the passing of the horse and the requirements of the motor indicated the need for a more level road, the abandonment of the severe camber on either side of the road, and the provision of super-elevation on road turns. Clearly, attention would have to be given to road foundations unless the heavier motor transport was to be prohibited.[23]

A new Road Board was established in 1910, and financed from the proceeds of a new motor tax. For the first time engineers were appointed to the staff of a central government body. The Chief Engineer of the Board, a former County Surveyor of Kent and his small staff, gave some limited support to their colleagues out in the counties, and the Board made small sums available for road improvement schemes undertaken by the counties. It did prepare some plans for new roads, but these plans were held back at the time pending a period of 'high unemployment' during which it was envisaged they could be used as public works to absorb surplus labour.[24]

The Board only lasted nine years. In 1919 a new Ministry of Transport was established and the first Minister appointed, Sir Eric Geddes - later to be immortalised in a different context. The professional presence was retained under a chief engineer, who was once again a former county surveyor, while the country was divided up into regions, each with a divisional road engineer to look after the central government's interest in roads. Again, almost all these engineers had previously worked in local government. Herbert Morrison, a former Minister of Transport, described the 'engineering staff of the Ministry acting as the helpful colleagues and big brothers of the hundred of municipal and county engineers rather than as superiors'.[25]

THE HIGHWAY ENGINEERS IN THE NINETEEN-THIRTIES

Between the Wars there were few major new road schemes, the Ministry and the counties were mainly involved in maintenance and improvement works. Nevertheless, during the thirties there was a surge of interest among the engineers in the idea of a major national system of trunk roads linking up the main centres of population. In 1930 the Institution of Highway Engineers was formed and, two years later, the British Road Federation was founded.[26] While in the mid thirties, the Government began to take the first hesitant steps towards creating a national road network. In 1936 the Trunk Roads Act made the Ministry of Transport the highway authority for certain specified roads which constituted a national system of routes. The construction, maintenance and improvement of these roads was to be carried out by the county councils acting as agents for the Ministry. In the same year, the newly formed Institution of Highway Engineers, which included central and local government engineers as well as people from the road construction industry, submitted a national highway scheme to the Minister. This scheme, clearly influenced by the new German autobahnen, proposed a massive construction effort involving the building of fifty-one new lengths of road totalling 2,826 miles. A year later, a large group of interested people visited the new autobahnen in Germany. Known as the German Roads Delegation, the group comprised 224 members who were drawn from Parliament, highway authorities, the engineering profession and others interested in the operation of vehicles.[27] The sheer size of the delegation and the range of interests represented indicates that the government was giving serious consideration to the idea of a motorway network on the German model.

Following the visit, the County Surveyors Society, which was well represented on the delegation, produced their own, more modest scheme for a thousand miles of motorway.[28] It has been later claimed by many county surveyors that this formed the original blueprint for the post-war roads programme.[29] This was forwarded to the Minister by the County Councils Association in 1938, presumably with the CSS's blessing. Soon after this, the Minister gave permission for work to begin on a scheme by Lancashire County Council to build sixty-one miles of experimental motorway; a scheme that was not to be completed for over twenty years.

The outbreak of the war led the government to shelve these plans. However, when attention began to turn to the tasks of post-war reconstruction, the blueprints began to reappear. In 1942 the Institution of Civil Engineers published a report advocating a thousand mile system along the lines previously advocated by the County Surveyors, and at an estimated cost of £60 million.[30] In the following year a Joint Committee of the CSS and the Institution of Municipal and County Engineers produced a report setting out possible design specifications for future motorways - recommending that they should be out in the country, away from housing development and giving design details, for example, the width of dual carriageways and the size of the hard shoulder.[31]

It seems hardly coincidental that, at the same time, similar plans were being prepared within the Ministry of Transport. Sir Frederick Cook, the Chief Engineer, wrote a Ministry memo on post-war planning which referred to the CSS proposals in stating the case for motorways. This became public in 1943 when Noel Baker announced in the House of Commons that the Ministry were preparing plans and, in the next year, a draft design paper on types of motorway interchanges was produced.[32] Also, in that following year, the Ministry Chief Engineer recommended a national motorway plan. Meanwhile the road construction industry began to prepare itself for an increase in the demand for road materials and in 1944 the Federation of Coated Macadam Industries was formed with the former Ministry Chief Engineer as Director.[33]

THE BEGINNINGS OF THE ROAD PROGRAMME

At first it appeared that the impetus for road building would be carried on into the post-war period. In 1946 Alfred Barnes, the Minister of Transport in the new Labour Government, announced a ten year programme of trunk road construction. This programme closely resembled the CSS proposals, involving as it did the construction of a thousand mile network linking up the major industrial and commercial centres.[34] In the same year there was another visit to Germany, which included local and central government engineers, to assess the post-war condition of Germany's highways and to learn more of the autobahn system. The Report of this second delegation noted some of the advantages of motorways, stressing the enhanced safety obtained from separating traffic and pedestrians, from designed access points and from the uniformity of design.[35] After this visit in 1949, legislation was passed to change the regulations allowing motorways to be built.[36]

Despite these early signs of interest from the government - and the activities of the roads lobby - motorway building did not begin in earnest until the mid nineteen fifties. Instead, the resources available for post-war reconstruction went into other, more politically salient, policy sectors like housing. So, between 1948 and 1953, the average annual expenditure on minor improvements and maintenance was, in real terms, 66% less than the average between 1936 and 1938, while expenditure on new construction and major improvements was a mere 21% of that between 1936 and 1938.[37] But during these years car ownership remained at a relatively low level, while the bulk of commercial goods was still carried by the railways. As yet the pressure from a wider grouping of road interests on government was only beginning to be felt.

The turning point was in the mid nineteen fifties. By 1960 annual expenditure on roads had risen to £55 million, and work was beginning on a new national system of trunk routes, with the Ministry taking overall responsibility and the counties undertaking the construction as agents for the Ministry. The first length of motorway - the Preston Bypass, first planned in the thirties - was completed in 1958, designed and built by one of the best known of county surveyors, James Drake, now Sir James Drake.[38] Unfortunately, it was closed just three weeks later after the surface began to disintegrate.[39] In the next year, the first 73 miles of the M1 were opened. But again this road was beset with design problems and the road surface began to show signs of strain under traffic stress.[40] This was partly the result of inexperience in the use of materials under conditions of high speed and heavy loads, plus underestimates of the amount of traffic that the road would have to carry. Nevertheless, despite these early setbacks, by the early 1960s new road building had begun on a large scale.

It is not easy to discern the major reasons for the beginning of a large scale roads programme in the mid nineteen fifties - in particular to draw any conclusions in respect of the relative influence of the roads lobby and the professional community. Certainly during this period the roads lobby was very active. Three of the major interested parties - the British Roads Federation, the Institution of Highway Engineers and the Society of Motor Manufacturers and Traders - formed a joint committee and presented the Minister of Transport with a Report outlining what they considered to be the main economic benefits of roads.[41] The County Surveyors Society, the majority of whose members also belonged to the Institution of Highway Engineers, 'produced covering memoranda, and the name of Graham Bennett, County Surveyor of Oxfordshire (afterwards Bucks), should be particularly remembered as the spearpoint of the pressure'.[42] The road construction industry, as well as being represented in the Institution of Highway Engineers, was also organised by this time into a number of trade associations, such as the Federation of Coated Macadam Industries.[43] The growing road haulage industry brought in more voices to join the clamour for new and improved roads. Finally, as car ownership began to grow, following post-war prosperity, the political importance of private motorists increased, as reflected in the growing membership of the Automobile Association and the Royal Automobile Club. Whether the tactics of these groups were decisive is difficult to say. In one account of the roads lobby during this period, Finer expresses doubts about the potency of the roads lobby. His conclusion was:

> What else <u>can</u> a government do but plan new roads? The alternative would be to restrict private motoring, and that is, politically, quite impracticable.[44]

This is not so much an explanation of how the government became committed to road building as a statement of the prevalent assumption at the time.[45] The assumption was that the response to the growth in traffic should be simply to build more roads and improve existing ones. And the Ministry of Transport took 'great pains to elaborate on and promulgate the basic and simple argument that roads were needed to relieve traffic congestion, which would get worse as traffic grew'.[46] So, although the early response to the growth of motor vehicles had been to try to restrict their use in order to limit their impact on the roads, by the nineteen fifties the response had changed to building roads to anticipate the likely expansion in traffic. This new assumption was not seriously questioned for almost twenty years.

There was a further aspect to the growth of the roads programme, which is often neglected in the stress placed upon private car ownership, and that is the growth in commercial vehicles that occurred from the middle fifties onward. Not only did the road hauliers themselves have a strong interest in improved roads, but also much of private industry which stood to profit from cheaper and faster road transport. Moreover, it became widely assumed that the economic growth of an area depended, in part, on its connections with a national motorway network. This meant that many of the 'depressed regions' pressed for new roads, assisted by the influential groups of Labour MPs representing these regions. Against this background, then, the bipartisan consensus that arose within Parliament and the country, favouring the construction of roads, can be explained. This consensus only began to break up in the seventies as, on one hand, environmental concerns came to the forefront and, on the other, many within the Labour Party began to define the central transport problem in terms of social welfare and need rather than in terms of responding to demand. But before these changing problem definitions are discussed, the development of the strong policy community, centred on roads and dominated by professional engineers, will be discussed.

THE CHANGING PLACE OF THE ENGINEER IN CENTRAL GOVERNMENT

While there had been engineers in the Ministry of Transport since its foundation in 1919, their numbers had only grown slowly over the years. Until 1964, England and Wales had been divided up into ten engineering divisions, each headed by a divisional road engineer, while in the Ministry headquarters there was, under a deputy secretary, a Chief Engineer with six engineering divisions under assistant chief engineers, and ten further divisions of administrative and executive staff. But the upsurge in roads expenditure in the early sixties also brought organisational changes in the Ministry. The first recommendations for changes were made in a 1962 Report of the House of Commons Expenditure Committee.[47] In their view the Ministry engineers were too involved in checking and examining the work of the local authority engineers and so they recommended a review of the use of engineering staff in the Ministry. This Report, and the organisational problems involved in implementing the expanding roads programme, led the government in 1962, to set up a review of the Ministry's organisation. The remit of the Committee appointed was not confined to the relationship between local and central government engineers but included the relationship between the DREs and headquarters, and that between the professional and non-professional civil servants.[48] The subsequent Report, the Heaton Report, rejected the Estimates Committee's charge of excessive supervision of local authorities by the Ministry engineers. Instead, they recommended that more, rather than less, guidance could usefully be given to local authorities - a conclusion perhaps not surprising as the Committee consisted of civil servants with only one outsider. But the civil servants' real concern was with the internal organisation of the Department and they were unwilling to relinquish control over local authorities. The Report recommended more devolution within the Department, with the divisional road engineers being given more delegated powers to deal with issues without reference back to headquarters. They also wanted to see a new post of Director of Highway Engineering created as the joint head of the Ministry's highways organisation, with a deputy director from the administrative side. Further down the hierarchy as well the engineers were to be brought closer to the policy side of the Ministry. Some divisions were to be integrated with both a professional head and an administrative head, being jointly headed by assistant chief engineers and assistant secretaries.

These changes were implemented soon after the Report was published. They were seen within the Ministry as improving the morale of the engineering staff, raising the standing of the Ministry's engineers in the eyes of the local authority engineering staffs, and increasing the efficiency of the roads organisation by bringing the policy and advisory functions closer together.[49] The changes had considerable significance for the standing and the influence of the engineers within the Ministry. They now became increasingly active participants in the making of roads policy rather than simply fulfilling an advisory role.

At the same time professional engineers, through their involvement in administration in the divisions, also gained more autonomy. The fact should not be overlooked, either, that the Ministry's engineers had, in partnership with the administrative side, successfully countered the Estimate Committee's recommendations that more freedom should be permitted to the local authorities.

The two internal reorganisations broadened out the Departmental involvement of the engineers beyond narrowly defined engineering questions. They began to become much more aware of the financial and political considerations behind the roads programme and so more able to temper their advice to increase its persuasiveness for the policy makers. It meant, too, that they had more frequent and informal contact with the administrators in the Ministry. In summary:

> The general administrator can take little comfort from the structural changes in the Ministry of Transport and other Departments. They all tend to extend to the professional civil servant responsibilities and duties which were previously reserved for the administrative (and sometimes executive) class civil servant.[50]

The headquarters organisation from this time remained much the same until the mid nineteen seventies. The one major change that took place was the creation of the roads construction units in 1967 and 1968. Most of the new construction and other road works were carried out by the county councils as agents for the Ministry, under the supervision of the divisional road engineers. This system of road building had emerged after the Trunk Roads Act 1936 which had made the Ministry responsible for trunk roads, previously the responsibility of the counties. The system had worked well enough when the construction of new roads remained at a low level but there was a feeling in the Ministry that it was not appropriate for dealing with larger and more complex projects. In particular, it was doubted whether the 'most efficient use of the scarce engineering resources available' was being made.[51]

In 1966 consultations over possible new administrative arrangements to deal with the enlarged roads programme began. At this time the County Councils Association, representing those councils most affected by any proposed changes, accepted the Government's arguments for larger units and increased involvement in the planning of trunk roads, but the association favoured the attachment of any new organisational units to the existing divisional road engineer organisation within the Ministry.[52] The County Surveyors Society took the same line and it would seem that the CCA position reflected that of the Surveyors. The Surveyors wanted to see change and had made their views widely known, 'we had moaned on for years and years about getting a proper basis for road construction, to escape from the rat-run of trying to get decisions out of the Ministry'.[53] However, the CCA and the County Surveyors wanted to continue to operate as agents for the Ministry but within an improved system of project planning under the divisional road engineers. Clearly they saw their interests as being best served if decision-making were decentralised within the Ministry itself and brought closer to the counties.[54]

The counties were not the only interest group pressing for changes in the organisation of road construction. The contractors and road material suppliers were also anxious for changes, as their interests lay in creating more secure and predictable markets for their goods and services. The new Labour Government had given these interests an ideal opportunity to express their views and press for action in the form of the new National Economic Development Council and its constituent committees. In 1965 one of the first acts of the Economic Development Council for Civil Engineering was to set up a working party into road construction, chaired by an industrialist. The priority that NEDO gave to road construction indicates the increasing importance given to roads by industry. The working party produced two sets of recommendations in 1966 and in 1967[55] and these recommendations were endorsed by Barbara Castle, the then Minister of Transport. She agreed that the Ministry should issue forecasts of work every quarter to indicate the next two years programme of work, and that the Ministry should allocate contracts in such a way as to enable certain firms to build up expertise in particular areas of work.[56] After 1966, too, there were regular meetings between the Ministry and representatives of the industry along the lines recommended by the working party.[57]

There were, then, strong pressures from industrial and local government interests for the Ministry to come up with a more programmed method of planning road construction. In many ways industry and local government had similar aims related to reducing uncertainty in order to enable them to plan ahead. In the event the Ministry's solution was probably more to the taste of the industrialists than to that of the counties. The Ministry proposed the creation of a new type of organisation between the central and the local levels and, in September 1966, the County Councils Association was persuaded to agree in principle to what later became the road construction units.[58]

It is difficult to say where this idea originated and why it should have been thought necessary to set up another roads organisation in parallel to the existing regional organisation of the Ministry. Certainly the county surveyors were hostile to the idea. One surveyor thought the units were 'just a gimmick', the sort of 'flashy idea' that appealed to Barbara Castle.[59] Many surveyors perceived the new units as a threat to local independence. One of the leading surveyors declared that he was 'not going to sell his birthright for a mess of potage'.[60] The surveyors felt that they had worked well with the DREs up until then and wanted them strengthened rather than, what they saw as, the further centralisation represented by the RCUs.

The impact of the new units on the surveyors as a professional community was divisive, at least at first. Whereas previously trunk road construction had been spread among a large number of authorities, under the new system construction was concentrated in just those sixteen counties designated as sub-units with the county surveyor as the sub-unit chief. Thus some counties lost construction work which was redirected towards the sub-units. Many of the county surveyors who did not come to participate in the RCUs were, at least initially, resentful of those that had enlarged organisations within their counties.

Appropriately, the first RCU was established in the area where the first motorway had been built, the North West. The unit was set up in April 1967, and the man primarily responsible for that motorway, James

Drake, the County Surveyor of Lancashire, was appointed to head the Unit.[61] The Midlands and North Eastern Units followed later that year and the remaining units were established early in 1968. The RCUs were, in principle, operated as partnerships between the Ministry and the sixteen 'participating counties', which had been selected by the Ministry. The participating counties contained the sub-units which were responsible for actually planning and supervising the construction of new trunk roads and major improvement schemes. This inevitably gave the county surveyors from these counties considerable influence in the formulation of the roads programme.

The six new RCU headquarters took over the responsibility for the construction of major trunk roads which had previously rested with the DREs and the Ministry headquarters. This left the DREs with smaller trunk road schemes,[63] and with maintenance; the DREs continued to use the counties as agents on both these types of work. The RCU headquarters, then, dealt with the administration of large schemes and with the major engineering aspects of new roads. The administrative work involved preparations for acquiring land and for public consultation and inquiries, and arranging contracts and financial details. Meanwhile, on the engineering side, the headquarters acted as the consulting engineers for the Ministry which was, in a sense, their client. These arrangements involved a great deal of delegation from the Ministry to the RCUs. That latter oversaw the evaluation of alternative routes, decided on which routes the public and the Landscape Advisory Committee should be consulted, advised on the choice of a preferred route, supervised the detailed design of roads within the sub-units, presented the Ministry's case at public inquiries and, finally, supervised the construction of roads.[64] The RCU headquarters acted as employer and engineer under contract for schemes prepared by the sub-units and as employer for schemes prepared by consultants. Meanwhile, the sub-units, based in the counties, prepared the road scheme in detail. They carried out the technical appraisal and evaluation of alternative routes, and then designed the preferred route in detail. During the actual construction, the sub-units supervised the works, providing the resident engineers and the permanent site staff. It should be noted that the sub-units worked in neighbouring counties as well as in their own home counties.

The staffs of the six RCU headquarters and their sub-units were a mixture of civil servants and local government officers. However, the senior headquarters staff - the directors, their deputies and the senior administrative staff - were all civil servants. The director of each Unit reported directly to the Director General (Highways),[65] and the directors were appointed at the same level as the Chief Highway Engineer. But the majority of RCU staff, both engineering and administrative, were local government officers. In 1979, for example, 81% of the RCU headquarters staff were from local government and 96% of the staff in the sub-units.[66] Most of the civil service staff in the sub-units were junior engineers who were acquiring the necessary experience to gain their professional qualifications.[67] In this way the RCU set-up provided a valuable function in enabling junior civil service engineers to gain practical experience, which they might not otherwise have been able to, as the Ministry was not directly involved in construction. This was quite an important consideration in the eyes of the Ministry engineers; its significance having already been highlighted in the Heaton Report, where such experience was seen as important in improving the morale of civil service engineers and their standing in the eyes of their local government counterparts. Certainly engineers characteristically stress the value of practical experience and assess each other in terms of the major construction projects on which they have worked. This suggests that the civil service engineers often experienced difficulties in legitimating themselves vis-a-vis their local authority colleagues, most of whom would inevitably have had more practical engineering experience.

THE EMERGENCE OF CHALLENGES TO THE ROADS PROGRAMME

The early 1970s were the boom period for road construction and for the highway engineers. The cumulative effects of the organisational changes within the Ministry of Transport, and later the Department of the Environment, were to create a strong professional community of 'roads men'. The community encompassed the Department and extended down to the county level, where local government engineers were directly involved in implementing the roads programme through the RCUs. Through the RCU organisation the engineering organisations of central and local government became closely knitted together - though a significant number of engineers in the counties remained outside the RCU organisation. Some senior local government engineers made the transition from local government into the civil service through the RCUs, men like James Drake, Maurice Milne (a former County Surveyor of West Sussex and Secretary of the County Surveyors Society) and Ron Bridle, later to be the Chief Highway Engineer. The new RCU directorship posts offered county surveyors a new step in their career ladders. Moreover, the next generation of county surveyors, appointed during the seventies, were to come predominantly from those who had some experience in the RCUs. At the same time in the Ministry, the integration of the professional and administrative hierarchies increased the influence and the status of civil service engineers.

Nevertheless, ultimately the development and maintenance of this professional community rested on the central government's continuing commitment to major trunk road construction, a commitment which remained largely unquestioned in national politics until the mid nineteen seventies. Nevertheless, before this the roads men faced growing challenges to the implementation of their programme. These challenges originated primarily in those localities where roadbuilding had become an amenity issue. As this opposition grew and spread in different parts of the country, it had a crucial impact in slowing down the implementation of the national roads programme. With the result that while in 1958 it was estimated that schemes took a minimum of four years to be realised, by 1965 the figure had become seven years, by 1973 it was up to ten years and by 1978 fifteen years or more.[68] The organisational innovation of the RCUs, intended to reduce this slippage, had not affected this trend. This was, at least in part, because the major factor behind this slippage was the growth of the environmental movement - 'the "slippage" problem was changing from a "technical" one of lengthy procedures, and inefficiency in scheme preparation, to a practical one of the growth of intransigence'.[69] Moreover, as the growth in public expenditure slowed down, the roads men found it more difficult to defend their policy sector in the face of strong competition for resources as they were finishing the financial year with larger and larger underspends.

From the very first there had been statutory procedures allowing those affected by new road proposals to voice their grievances. The Ministry was obliged to consult local organisations, allow time for consultation and to hold an inquiry if a local authority requested or if the Minister considered it necessary.[70] The inquiries were conducted by independent inspectors whose reports were not binding on the Minister. This meant that the Minister was in the anomalous position of being judge and jury in his own case. Even once these procedures were completed, the Ministry had to go through a further process of compulsory purchase with a further possibility of a public inquiry. Clearly these procedures gave plenty of opportunity for objectors to question and delay road proposals. The longer the delay in construction, too, the greater the chance of new facts coming to light that involved changes to the proposals and yet more delays.[71]

Quite early on the Ministry had attempted to anticipate criticism by setting up the Landscape Advisory Committee in 1956. The Committee included representatives from the established amenity associations, like the Council for the Preservation of Rural England, as well as representatives from other government departments and professional associations. But the Ministry tended to approach environmental issues predominantly in terms of the aesthetics and design of the roads themselves.[72] This meant that the Committee, hamstrung by its terms of reference, generally limited itself to recommendations relating to the use of trees as screening, changes in banking and other cosmetic alterations to the basic plans.

By the late nineteen sixties, the Ministry of Transport was not the only government agency facing 'grassroots' opposition. The government was obliged to respond by taking on board some of the ideas relating to public participation. In the area of planning the Skeffington Committee[73] reported on ways of involving the public in the new system of development planning.[74] Similarly, over road construction, new measures were introduced to increase public involvement - ostensibly, at least.[75] The roads men were unhappy with some of these proposals, especially the Minister's commitment to the earlier publication of alternative plans, which they felt would further increase the delays.

In 1974, the roads men suffered a further setback when the RCUs were brought under the umbrella of the Regional Offices of the DoE. The intention was to ensure that roads proposals made greater allowance for the objectives of the regional plans, though the government, too, was becoming concerned at the poor public relations ability displayed by the RCUs.[76] So from then on, the directors of the RCUs were required to consult with the regional directors at nearly every stage of plan preparation and the latter were placed in charge of the political aspects of the RCUs' work.[77]

The ministerial concern with their public relations abilities and their own realisation of the need to counter their critics, led the highways people in the DoE to set up two working parties to look into ways by which opposition at public inquiries could be overcome. The first working party dealt with the 'Presentation of the DoE Case at Trunk Road Public Inquiries' (the Flowers Report) and the other with 'Relations between RCUs and the Public' (the Milne Report). It is worth examining the second of these reports as it gives a picture of how the roads men saw their opposition; in particular, their discomfort at being ridiculed in public inquiries by counsel and well-informed objectors. The view taken of objectors in the Report was quite uncompromising. Three types of objector were identified in the Report - the 'statutory objector', the 'objector affected by the proposals', and the more sinister 'objector with social or political motivation'.[78] The Report accepted the legitimacy of the first two types of objector but questioned that of the third type. What they found especially disturbing about the third type of objector was that they were visibly becoming better organised at both the regional and the national levels. The same people were being seen at more than one inquiry and, to the roads men, they appeared opposed to all new roads in principle:

> The disturbing feature is the emergence of organised bodies who have hidden motives that may have little or nothing to do with highway schemes under consideration. It would seem that a number of people have realised that the highway consultation procedures provide an admirable forum which can be exploited in a most satisfactory way from their standpoint.'[79]

Unfortunately, the Report did not go on to explain what this viewpoint was. But it did detail some of the RCUs' experiences of organised objectors, for example, sending 'astute probes' to the different agencies involved in the hope of being able to exploit unco-ordinated replies, saturating the public with their views through sympathetic journalists, adopting an 'emotive' approach,[80] and attempting to broaden out inquiries to consider overall highways policy.

The Chairman, Maurice Milne, and his colleagues came up with several recommendations intended to counter these tactics. In particular, they felt the need for a strong renewal of the political commitment to road construction, so they called for a 'firm ministerial policy statement';[81] while the Department itself had to put over a stronger case for roads and overcome some of the anti-roads arguments.[82] Within the RCUs they wanted to see improvements in the flow of information to the public and also a greater exchange of information among the units on the tactics employed by objectors, as a first step towards countering them. They felt, perhaps rather naively, that if the proposals were carefully explained all but the 'politically motivated' objectors would be persuaded of the Department's case.[83]

The Milne Report has been discussed at length as it provides an important window on the views of the roads men and their growing unease with the challenge from, what could be called, the amenity or environmental issue network. In the following months and years, this issue network became stronger and better organised. National pressure groups, like Friends of the Earth, became involved in the issue and a National Motorways Action Committee was set up in 1974, while local protesters began to attack the assumptions underlying the roads programme, challenging the need for a road rather than just arguing over its routing.[84] But 1976 proved a decisive year for the anti-roads protesters. In that year there were two major public inquiries into Departmental road proposals that ended in total disruption as protesters refused to permit the inquiries to proceed. Their central objection was that the inquiries were a denial of natural justice as the Department was judge and jury in its own case and, moreover, the Department refused to debate the road traffic predictions that formed the main argument for the road. As far as the Department was concerned, the Minister had agreed to the road and the inquiry was only there to advise him on the choice of route. The protesters, with their strong interest in preserving the environment, demanded that the inquiries should consider the necessity for the proposed road. In the Aire Valley Inquiry, the Inspector felt obliged to ask the police to remove protesters and to continue the Inquiry in private. Even so, the DoE had finally to abandon the Inquiry in the face of picketing protesters outside the inquiry hall.[85] Similarly, the Winchester M3 Inquiry had to be abandoned after the Inspector failed even to begin it in the face of vociferous protests that lasted over five days.[86]

These, and other less well known inquiries, were politically embarrassing in the country and, after the Winchester debacle in 1976, there was an adjournment debate in Parliament. These pressures led the government to set up two further reviews of inquiry procedures and of the methods of appraising major road schemes. Indeed, on one scale of success, the ability to stimulate the production of government reports, the issue network has a high score. The first review to report was the Leitch Committee, on the methods of appraising road schemes, which was set up in November 1977. The Leitch Report recommendations were essentially concerned with improving the Department's case for public consumption as well as raising the quality of its economic assessments of road schemes.[87] One of the more important recommendations was that a standing committee on trunk road assessment should be appointed as a continuing source of advice to the Department, with a much wider remit than that of the Landscape Advisory Committee which continued in existence. This Standing Committee was appointed by the Secretary of State in 1978 and included representatives from the amenity network.

The second policy review in this area was concerned with, what was, by this time, the almost perennial issue of inquiry procedures. The review was published in April 1978 and, again, it put forward several reforms in the direction of the protesters including, significantly, the recommendation that there should be more parliamentary debates on the roads programme.[88] However, by this time were other, more significant changes beginning to affect the roads programme. These were the beginning of a gradual financial rundown in the programme itself and the subsequent organisational changes which affected both the counties and the Department of Transport. These changes will be the subject of the rest of the chapter.

THE DECLINE OF THE ROADS PROGRAMME AND CLOSURE OF THE ROAD CONSTRUCTION UNITS

Despite the creation of the RCUs and the internal reoganisations of the Ministry, slippage remained a serious problem. Slippage was mainly attributed to the delays created by the public participation procedures and their supposed abuse by activists, but other problems relating to the scheduling of the construction also played a part. As Painter argues, the effect of this slippage was to make the roads programme:

> ...very vulnerable to cuts in a growing Whitehall mood of limiting expenditure growth. Within the DoE the inability of the roads men to spend their money made them easy prey for competing programme advocates.[89]

In 1978 the roads men got the type of ministerial statement for which they had asked in the Milne Report. But the statement did not accord with their own commitment to increased road expenditure. The Minister now broke with the earlier view that the roads programme was aimed at the completion of a total network of strategic roads. Instead, he announced the government's intention to adopt a 'more selective approach' to the future planning of road schemes. In the government's view, the major road network was nearing completion and the main schemes necessary to finish the network were already in the preparation pool. Accordingly a new set of criteria was drawn up to judge future additions to the preparation pool and review existing schemes already in the pool, but only 15 schemes out of 300 were dropped immediately as a result of this review.[90]

During the same period there was growing pressure from the private sector for changes in the organisation of road construction. Again the NEDO Committee on Civil Engineering was one vehicle through which this pressure was exerted. But this time the pressure came from the consulting engineers. They were concerned with the implications of the expansion of in-house engineering design teams in the public sector, especially in the road construction and water industries. They believed that they were losing work as a result and argued that it was necessary for them to have substantial practices at home if they were to continue to expand their overseas markets. But the 1977 Report of the NEDO Civil Engineering Committee did not support this view. Instead, it concluded that the needs of the consultants could be accommodated without the consultants taking over the bulk of the work from the public sector and found that it was generally cheaper to use the sub-units for construction than to use the consultants.[91] At this time too the local authority associations strongly rejected the suggestion that the private consultants should be allocated more road construction contracts; interestingly, the suggestion was made by a Labour Minister and rejected by a Conservative ACC.[92]

It was after the announcement of the revised roads programme in 1978, that the government set in motion a review of the RCU organisation, for the decline in the roads programme had thrown into doubt the continuing need for the RCUs. Already by 1978, the actual expenditure on road construction was only half the level that it had been five years before, and a third of what it had been ten years before, when the RCUs had been established.[93] Allowances were already being made, in the agreements between the participating counties and the Department of Transport, for falling workloads and the likelihood of closing some or all of the RCUs and their sub-units. So, it came as no surprise when, in July 1978, the Department of Transport issued a consultative paper on the rundown of the RCUs.[94] This paper recognised the implications of the 1978 Roads White Paper and maintained that any new form of roads organisation had to take account of a continuing fall in additions to the road schemes preparation pool, and the demand from the counties for the inclusion of more schemes of local significance. And the White Paper too envisaged that the trunk roads rundown would mean more resources for local roads - a development welcomed by the counties.

The major problem facing the government was what to do about the staff of the RCUs. At this time there were 2,500 officers in the RCUs, just 200 of them civil servants while the rest were local government officers. Nevertheless, the Secretary of State was directly responsible for these officers in his role as employer, as originally negotiated in the sub-unit agreements. The Government inclined to the view that the 'most flexible situation might be to revert to the situation where most of the design work is unambiguously associated with local government'.[95] The position, then, would be a return to the pre-1967 arrangement, whereby counties would implement most trunk road schemes - though the Department would retain the right to determine whether a scheme was implemented on an agency or other basis. Clearly the intention of the government was to transfer the responsibility for the local government staff back to the counties.

This statement of Government intent was in line with the views of both the Association of Metropolitan Authorities and the Association of County Councils. The ACC had been unhappy with the RCU organisation for a long time and they seized on the falling roads programme as supporting the case for the counties ultimately to take over responsibility for trunk roads. At the time of the establishment of the RCUs the counties had opposed them, as they would have preferred that the then divisional road engineers be given the responsibilities in the implementation of the roads programme which went to the RCUs. But the participating counties were concerned that they might be lumbered with ex-RCU staff for whom they had no work. This problem was compounded by the reluctance of some county surveyors in the participating counties to lose their sub-units. In the negotiations that began with the Labour Government, and in the subsequent negotiations with the Conservative Government, the ACC based their case on three main arguments. Firstly, they argued that since the counties were charged with the legal responsibility for formulating strategic planning policies, in which transportation was clearly an essential factor, and since they were also responsible for co-ordinating public transport, the counties were in an ideal position to take over increased responsibility for trunk roads in their areas. Secondly, if the emphasis was now on small, local schemes then the counties were in an ideal position to take on responsibility rather than the DTp. Thirdly, the RCU and sub-unit staff were predominantly local government staff and, therefore, in part at least, already the responsibility of the counties.[96]

The associations expected the government to set up a joint DTp-associations working party, as was conventional on these sorts of issue. Instead, a group of DTp civil servants toured around the counties to study the situation and, then, the Department began negotiations over the future of the RCUs on a regional level, so circumventing the associations. At the same time, negotiations were started with the staff sides in local government and in the Department. The local government officers were represented by the Association of Public Service Professional Engineers (APSPE)[97] and the National and Local Government Officers Association (NALGO).

During the autumn of 1978, the Department asked for comments on a further consultation paper setting out the options for RCU staff.[98] The paper was directed towards the union side. It referred to the problems that arose from the 'third force' position of the RCU staff, because their 'role in serving the Secretary of State's interests requires them to stand somewhat aside from their employing county'.[99] Many of these staff had never worked in local government in any case, and many of them, the paper continued, faced a bleak future as they could not easily be accommodated within the counties, and those able to find new jobs had already been moving out of the units. This latter problem - the loss of many qualified and key staff - was making the resolution of the RCUs' future more urgent. The three options outlined in the consultation paper were: to continue to make piecemeal arrangements for staff as work was terminated; to bring the RCUs into the civil service by offering them a chance to compete for civil service entry; or, to simply wait and see.

The local government staff side, APSPE and NALGO, favoured the slow rundown option. But they wanted guarantees that there would be no compulsory redundancies, that officers would not be required to move around the country without compensation and that they would have preferential consideration if they applied to join the civil service.[100] But the Whitley Council Staff Side in the Departments of the Environment and Transport were quite opposed to the recruitment of local government officers into the Departments. They considered that this would further reduce the opportunities of promotion for civil service engineers and were not prepared to see local government engineers appointed to senior posts.

Negotiations continued with the counties and the staff unions during 1978 and into 1979. The RCU headquarters staff posed the most difficult problem but agreement over the rundown of the sub-units was easier. Generally speaking, the proposals were that the sub-unit staff and workloads would be spread among more counties, with sub-unit schemes in non-participating counties transferred to those counties as the agents. The sub-unit counties would then revert to agency agreements, thus enabling their staff to transfer to county schemes. By early 1979 the counties, on a regional level, were close to agreeing with DTp arrangements for this transfer to take place.[101]

THE PRIVATISATION OF TRUNK ROAD CONSTRUCTION

Final agreement on the Labour Government's proposals was not reached before the Conservative Government entered office in May 1979. The new Conservative Secretary of State, Norman Fowler, froze the negotiations. In August 1979 the Department announced a new review of the workings of the RCUs and of possible changes in the organisation of road building. The review was conducted under the auspices of Sir Derek Rayner, a businessman who had been brought in by the Conservatives to promote efficiency and to root out waste in central government.[102] Why such a review should have been done in consultation with Rayner is not clear. Cynically, it might be supposed that the government expected him to come up with their own favoured solution - that the work should be given to private consultants - or it could indicate some Conservative suspicion of any negotiations dominated by civil servants and local government officers. Up to this point it was the latter who had actually conducted the negotiations. However, the Rayner study was based on research conducted by a civil service principal attached to Rayner. At the same time consultations were begun again with the staff unions, the local authority associations and other interested parties, including the Association of Consulting Engineers.

The Rayner Report on the RCUs was published in February 1980 and, not unexpectedly, it came down in favour of private consultants rather than the counties. The Report shared the earlier Labour Administration's concern over the continuing viability of the sub-units once their demise was certain - morale would fall further and natural wastage, already running at 20% a year, would increase.[103] Uncontrolled natural wastage could rapidly reduce the viability of a sub-unit, perhaps preventing it from completing schemes. But, whereas the Labour proposals had envisaged counties as agents taking over schemes and excess staff being allocated to county road schemes, Rayner argued that this assumed that expenditure on county road construction would be increased. If this assumption did not hold, he claimed that there would be surplus staff in the counties. The Report went on to look at the different performances of sub-units and consultants in the light of the earlier NEDO Report. It rejected the earlier Report's conclusion that the sub-units were cheaper than consultants and asserted that the cost differences were negligible. Rayner maintained that, in the changing conditions, the advantages were increasingly with the consultants: 'In the circumstances of a declining workload and of constraints on public spending, it is a significant advantage of consultants that they are able to offer the Department a staff resource which is flexible and not self-perpetuating'.[104]

One of the other major advantages of consultants mentioned in the Report was that it would enhance central control - both over the finances[105] and over the projects themselves. It was suggested that the counties had been allowing local considerations to influence their allocation of priority among projects, and so did not necessarily reflect national priorities.[106] This probably reflected the new Conservative Minister's concern that the Department's intentions were not always fully respected by the agent authorities. This view too is in line with the distrust of local autonomy which has become characteristic of this Conservative Government. Of course, removing the counties from an involvement in trunk road construction reduced one source of pressure on government roads policy. It should be added, though it is a less important point, that many of the DTp engineers themselves felt that they could exercise more influence over private consultants that they could over the counties, even given tightened up agency agreements.[107]

After the publication of the Rayner Report, the government moved rapidly in producing a set of 'Ground Rules for an Action Study' of the RCU rundown. These rules stated that there was to be a presumption in favour of consultants so that the onus would be on the counties to make out a 'special case' to retain even those existing schemes on which their staff were already working. While, in the case of new schemes, counties would have to demonstrate that there was a 'public advantage' in their managing a scheme, rather than private consultants.[108]

The ACC and the County Surveyors Society reacted angrily to these 'Ground Rules'. The ACC sent a delegation to the Secretary of State to protest at the strong bias, which favoured private consultants, against the county councils. The counties were only willing to accept equal competition between themselves and the consultants on a 'value for money' basis. But the Minister made it clear to the Association that the government was committed to the objective of getting road construction out of the public sector and into the private sector - almost without regard to cost differences. The arguments advanced for this objective were that it would enable greater flexibility in the use of staff - that is reduce the resistance of staff to moving - and that it would strengthen the home base of UK consulting engineers.[109]

The County Surveyors Society had, in part, helped to stimulate the ACC reaction. They had presented a paper to the ACC Planning and Transportation Committee attacking the the 'Ground Rules'. This strongly worded attack expressed the Surveyors' fears that the Rules would 'further erode the position of county councils as competent authorities to take over schemes on an agency basis'.[110] They also feared that it would reduce their, and the counties', ability to influence road schemes:

> If work is put out to consultants and directed by DTp, the ability to influence schemes in the formative stages will be much more difficult. At the very least they are going to lose touch with what is going on.[111]

They went so far as to suggest that the proposals could be the prelude to 'something more sinister regarding counties' long term future'.[112] In any case, the Society argued, few consultants would have the necessary staff already in post and they saw difficulties arising from the lack of continuity in transferring schemes. On new schemes, the presumption in favour of private consultants also appeared unfair to them. Finally, they were angered that consultants were going to be offered the smaller schemes which had previously been the counties' sole preserve. The ACC noted the comments of the Society but took the line that the government's proposals were not unfair to the staff and not wasteful of public funds. The final view of the Association was that the uncertainties consequent on further delays made further opposition to the government counter-productive. In any case, the government were clearly not going to be deflected in pursuing an object with which many of the elected members on the ACC committees sympathised. There was a feeling around too, and this will be illustrated in the chapter on highway agencies, that Conservative local authority associations should not be seen to be in sustained conflict with a Conservative Government.[113]

Although the ACC had finally agreed to the government's proposals, the Association did not abandon their long standing commitment to press for increased trunk roads responsibility for the counties. In July 1980 they were able to put their case. The opportunity arose in the form of the Parliamentary Select Committee on Transport, which had elected to consider the Roads Programme in the light of the June 1980 Roads White Paper.[114] The Association submitted written evidence and the Chairman of the Planning and Transportation Committee, one other elected member, the Secretary of the County Surveyors Society and an ACC Under-Secretary, attended to give evidence as witnesses. In their written evidence, the Association agreed with the rather general priorities identified in the Roads White Paper - the necessity of tailoring roads policies to available resources and to the encouragement of economic development. The stress on new bypasses was also welcomed as contributing 'great benefit to historic towns and rural communities'. What concerned them, in line with the earlier CSS Report, was the decision by the government to defer the design work on over a hundred new schemes. For the counties, this had immediate staff implications as well as increasing the problem of future slippage, for it meant that there were fewer schemes in the preparation pool to be brought forward. But what they really wanted to see was extensive de-trunking, with the counties taking over responsibility for more roads from the DTp. Their argument was that 'the county councils are the best people to take greater responsibility in the road programme and less should be done by the Ministry of Transport'.[115]

Similarly, the AMA, in their evidence to the Select Committee, concurred with the ACC in advocating that the counties should be given further trunk road responsibilities. In fact, the AMA went so far as to question the continuing need for a national trunk road authority. It seemed to the AMA that the metropolitan counties, given their expertise in engineering and their role as strategic comprehensive transport planning authorities, were ideally placed to take over responsibility for trunk roads. But, in contrast to the ACC, they took a less optimistic view of the government's intention to shift future resources towards county roads. In their written submission, the Association maintained that the White Paper actually implied a 'radical redirection of funds away from local transport capital'.[116]

The Select Committee Report, published in November 1980, supported the views of the counties. While agreeing with the government's declared intention to direct funds towards relatively localised schemes, it drew two opposite conclusions from those of the government.[117] The Committee felt that this change in emphasis in the roads programme clearly implied that there should be more, not less, local involvement. In the long term, their view was that the Department's highway functions should be reduced - in particular, removing their direct involvement in the design, construction and maintenance of highways. Moreover, the Select Committee considered that the Department had not taken sufficient notice of local, county views in the Roads White Paper - although the DTp, in their evidence, had assured them that the Department took notice of the county structure plans and Transport Plans (TPPs) in drawing up the lists of schemes. The Committee concluded by calling on the counties to do some detailed work on the ways in which they thought that such a transfer of functions could take place and recommended that an independent inquiry should examine possible arrangements for roads responsibilities after the completion of the motorway programme. Significantly, too, the Committee regretted that Parliament had, in the past, not taken an adequate interest in roads policy - apparently reflecting the widespread concern in the 'anti-roads' lobby that roads policies had never been properly debated in Parliament. Accordingly, they recommended that future roads white papers should be the subject of a full debate in the House.

The government continued with their plans to transfer the bulk of road construction to private consultants. In July 1980 the Department set up a 'clearing house' to look after the interests of those staff whose schemes were being transferred to the counties or to the consultants.[118] In the October, the Secretary of State announced the government's final intentions.[119] Over three-quarters of the work then being handled by RCU sub-units was now intended for consultants, who were given a month in which to submit tenders. According to the Minister, many consulting firms had expressed an interest in taking on staff from the RCU sub-units; something was clearly in the firms' interests, especially where the schemes were already under way. But the DTp was only willing to take two hundred extra staff into the DTp and these would have to apply for the posts through the usual recruitment channels.

The ACC expressed alarm at the speed of the handover to consultants. They accused Norman Fowler of indulging in 'speed for the sake of speed'. The Chairman of the ACC Planning and Transportation Committee asked that the counties should be allocated more work during the transition period, while the 'consultants geared themselves up to absorb available staff and work out how they were going to make charges for their work'.[120]

Unusually, the Institution of Municipal Engineers entered the lists against the government. Their President, incidentally a metropolitan county surveyor, wrote to the Secretary of State protesting that the 'traditionally balanced treatment of agent counties and consultants' had shifted too far in favour of the latter.[121] Furthermore, expert teams of engineers which had taken years to build up within the counties were being broken up as a result of the government's policy. But the Secretary of State replied that the traditional balance was no longer appropriate, besides which, he did not intend to break up these teams but simply to transfer them to the private sector. He added the blunt comment, 'It is not for me to comment on municipal engineering as a profession, but I do think that consultants are likely to be able to provide better and certainly more varied future prospects for staff than the public sector'.[122]

In December 1980, Norman Fowler confirmed the earlier proposals in the House in response to an inspired parliamentary question. The President of the Municipals expressed his disappointment at the final outcome:

> Clearly, a political decision has been made to transfer work to consultants - we can only accept it, although I still believe that there is a need for balance in the way work is allocated, It would be very unfortunate if in future, every political change brings a big swing of work - from consultants to the public sector and back to the consultants. It's not in anyone's best interests.[123]

So by this time the local government engineers had to accept the loss of work to the private sector. But the County Surveyors Society was determined to continue to press for counties to take over trunk roads, encouraged by the Select Committee Report. Immediately after the publication of the Select Committee Report, the ACC set up an officer working party to examine the possibility of transferring responsibility for trunk roads to the counties in line with the Committee's recommendation. The CSS was, of course, well represented on this working party, though it did not quite hold a majority.[124] The ACC approved the Report (though it is not known how or whether they amended it) and it was released in July 1981. The Report found that there was 'tremendous potential for economies in public expenditure by transferring the executive and management functions of the Secretary of

State for Transport and his Department' to the counties.[125] The ACC wanted to see trunk roads transferred to the counties after three years,[126] though they considered that the Department should continue to be responsible for the motorways until the network was completed. The Association felt the Government could exercise adequate supervision through the TPP system. The 'only real obstacle to progress in this respect will lie in convincing those who have an interest in the status quo that there is a need for change and that the time for change is at hand'.[127] This passage was 'a clear attack on DTp civil servants' according to Municipal Engineering.[128]

However, this was not the only threat to the beleaguered engineers within the Department of Transport. In a report on the organisational structure of the DTp, commissioned by the Department, Dr Francis Walley recommended that the Department should loosen its supervision over bridge and road maintenance work performed by the counties. The Report wanted to see this supervision replaced by regular meetings between the counties and the DTp regional offices. Perhaps not surprisingly this Report has not been released by the Department, which appears to have been more than a little embarrassed by its findings.[129]

There was another serious move by the government, in the direction of further privatisation, in April 1981. The county surveyors had been worried that the privatisation of trunk road construction was the thin end of a wedge that had trunk road maintenance at the other end. These fears were confirmed when the Department announced its intention to conduct a study of the routine maintenance of trunk roads at the Standing Committee on Highway Maintenance.[130] The Department itself had been concerned for some time with the wide expenditure variations among the counties on maintenance expenditure.[131] But the obvious implication of this announcement was that the government were considering the privatisation of routine maintenance. Although this was just a study of routine and not structural maintenance, the county surveyors reacted angrily - 'we got on our high horse, no way were the counties going to execute this work'.[132] Ostensibly their anger was that the government was setting up the study unilaterally without going through the usual channels, and did not propose convening a joint Department-associations steering group - the convention in investigations like this involving both central and local government interests. The engineers obtained elected member support for their stance and asked for a further meeting with the Department. At the meeting the civil servants made it clear to the officers that the Minister saw the issue as the responsibility of the Department which was responsible for maintenance, and not the concern of the local authorities.[133] This left the local government engineers with little option other than to go along with the study and hope, as they believed to be the case, that it would be discovered that the counties were efficient. In July the Department appointed a management consultancy company to examine routine maintenance, including the scope for privatisation.[134]

By the early nineteen eighties, highway maintenance had become the major concern of engineers in the public service. In part this was the result of the decline in the roads programme and the growing feeling among these engineers that their futures were less likely to be in what had, until then, been the glamorous and high status side of civil engineering, major road construction. But the county engineers began to use their claims to special knowledge of the problems of maintenance to argue that they should be able to influence the specifications given by the DTp to the private consulting engineers now involved in road construction.[135] This lever was quite powerful because of the continuing repair and maintenance problems on trunk roads and, in particular, on motorways.[136] There had already been growing government concern, reflected in increased allocations to motorway maintenance.[137] This government concern is perhaps a measure of the effectiveness of the civil service engineers in convincing the politicians that there was indeed a crisis in the condition of many of the country's motorways, though the DTp engineers continued to insist that more funds should be made available.[138]

CONCLUSION

From the first, municipal engineering was involved implementing ambitious programmes of, initially improvements to public sanitation and, later on, extensive improvements to the roads. These new tasks were large in scale and unambiguously seen as falling within the province of engineers, and there were, of course, no other occupational groups in situ to challenge the engineers. Undoubtedly the fact that the engineers were the first 'technical' profession in local government was an important advantage for them over the later professions such as architecture, an advantage that has persisted into more recent times. Municipal engineering in the nineteenth century also had strong reforming associations which gave it a sense of mission then, which it has only partly maintained as its tasks have ceased to be centred on public health and the campaign against disease, and become increasingly technobureaucratic. At the time this original mission enhanced the standing of municipal engineering and, in providing a sense of purpose, cemented the members of the profession together.

Of course, the engineers had the tremendous advantage of central recognition and support from the start. This support continued into the twentieth century when local authority engineers were closely involved in the new administrative arrangements to improve the country's roads and they were recruited to staff the new Ministry of Transport. The main role of these new civil service engineers was to encourage local highway authorities to improve their roads through setting a professional example, rather than through the application of material incentives or sanctions. In this way professionalism was used by the centre as a policy instrument in its response to the increasing public demand for better roads. By the end of the second World War, an influential professional lobby had grown up, straddling both local authorities, particularly the counties, and the Ministry of Transport. Influenced by the example of the German autobahnen, this lobby threw up a series of grand proposals for the construction of a network of new roads covering the whole country. The long delay between the first of these plans and the beginning of the roads programme only refelcts the strong competition for resources from other policy communities during the post-war reconstruction period, rather than any lack of influence by the engineers within their particular policy community.

At this time the County Surveyors Society became a major centre of influence within the closely overlapping professional and policy communities. As the roads programme expanded, the surveyors played a prominent role in contributing to the formulation of that programme and to its implementation. In fact the new road construction organisation gave them an opportunity to move on to senior positions in the civil service and the central administration of the roads programme. What emerged was a 'self-contained integrated hierarchy for road programme administration and technical work'.[139] This integrated hierarchy had its own strong sense of mission and organised itself against the growing opposition to the roads programme presented by the emergence of an anti-roads issue network - a network which was less well organised than the roads men supposed.

But this growing threat to the engineers' pre-eminence as the definers of the country's transport needs was accompanied by others at the local level. Apart from the threat of the environmentalist lobby, the county surveyors within their own authorities have experienced other pressures. The deference that the professions once enjoyed is no longer as automatic as it was. This changing level of public deference has become manifest in county surveyors' relations with the public and, more importantly, with their own elected members. Whereas, once 'the County Surveyor's word was that of a god',[140] during the seventies this became less and less true. By the mid seventies, the surveyors felt that 'the initiative was going away from us'.[141] This feeling led the Society to attempt to publicise their case in two successive reports on transport policy that took on board some of the new thinking that was around on transport matters, in particular public transport, while maintaining their emphasis on roads.[142]

Some mention should be made too of the county engineers' relations with other groups of public service engineers. Relations with one group, the civil service engineers, have not always been easy. One perennial problem has been centred on the local government engineers' complaints of excessive supervision by the civil servants. But the latter appear to have been successful in resisting any loosening of that supervision, even when such changes have been recommended by Parliamentary committees. More recently, the changes consequent on the rundown of the roads programme, have introduced new tensions into this relationship or, more accurately, highlighted tensions that were already there. Many county engineers have thought that their civil service colleagues preferred to use private consultants rather than counties in the form of agencies, mainly because the DTp can exercise closer control over consultants than it can over county councils - the Rayner Report on RCUs adds support to this view.

The other group of engineers which should be mentioned are those based in the districts. Local government reorganisation was a major setback to the district level engineers, especially those who had previously worked in the county boroughs. At reorganisation, highway functions were given to the counties, and water and sewerage functions to the newly formed water authorities. This left virtually no specifically engineering tasks at district level, consequently the district engineers have found it difficult to justify their continuing presence at that level.[143] Their main response has been to move increasingly in the technobureaucratic direction, becoming directors of technical services.[144] But highway agencies have proved to be an important source of tension between counties and districts as engineers in both types of authorities have fought to retain or regain highway responsibilities.

NOTES

[1] Edwin Chadwick, Report on the Sanitary Conditions of the Labouring Population of Great Britain, 1842, (ed) M W Finn, Edinburgh University Press, Edinburgh 1965

[2] Ibid.

[3] Ibid.

[4] Similarly there had been legislation thirteen years before, the Municipal Corporations Act, which was intended to remove the office of Town Clerk from the influence of corrupt practices and patronage, as well as raise the Clerk's level of competence. He was made the accounting officer for his corporation and effectively prevented from operating as a local politician. See: T E Hendrick, The Town Clerk in English Local Government, Allen & Unwin, London, 1960 p.20.

[5] Institution of Municipal Engineering, 'The Foundation of the Institution', (Chronology) p.1.

[6] This and the rest of the paragraph draw on IME, 'The Foundation of the Institution'.

[7] Ibid.

[8] The objects were given as:
a. The promotion and interchange among its members of that species of knowledge and practice which falls within the department of an Engineer or Surveyor engaged in the discharge of the duties imposed by the Public Health, Local Government and other Sanitary Acts.
b. The promotion of the professional interests of the Members.
c. The general promotion of sanitary science.
Ibid.

[9] Sidney and Beatrice Webb, English Local Government: The Story of the King's Highway, Longmans, Green & Co, London, 1913.

[10] The Institution of Civil Engineers was founded in 1818 and Telford became President in 1820. But the first civil engineering degree course was not set up until 1838. W.H.G Armytage, A Social History of Engineering, Faber & Faber, London, 1961 pp.22, 149.

[11] Webb , English Local Government, p.166.

[12] Ibid.

[13] The full definition of an adequate road surface, according to Macadam, was 'an artificial flooring, forming a strong, smooth, solid surface, at once capable of carrying great weight, and over which carriages may pass without meeting any impediment'. Quoted in Webbs, English Local Government, pp.172-173. It

should be added that this is what is meant by a 'macadamised' surface. Although Macadam is chiefly associated with the term 'tarmacadam', the use of tar as a road surface was not applied until several years after his death.

[14] Ibid, p.173.

[15] Ibid, p.173.

[16] Ibid, p.174.

[17] Thus it was formed 'to watch over all matters affecting the office of county surveyor' and 'especially to watch the interests of the Society with regard to any Bill in Parliament'. Quoted in Bill Harrison, 'Surveyors', County Councils Gazette, May 1977, p.44.

[18] Membership was now extended to 'civil engineers or surveyors holding chief permanent appointments under any municipal corporation, county council, or urban or rural sanitary authority of the like nature within the United Kingdom or in the colonies or in foreign countries'. IME, 'The Foundation of the Institution', p.3.

[19] Bryan Keith Lucas, The Unreformed Local Government System, Croom Helm, London, 1979.

[20] Webbs, English Local Government, p.248.

[21] Webbs, English Local Government, p.248. Mick Hamer notes that the association came 'to be dominated by tar interests. W.Rees Jefferys became secretary in 1901 and changes in the constitution marked its takeover by modern road construction and motor interests'. Wheels Within Wheels, Friends of the Earth, London, 1974, Appendix 1. The RIA is now effectively defunct.

[22] Webbs, English Local Government, p.245. and Herbert Morrison, 'Highways and Transportation', in Harold Laski, W Ivor Jennings and William A Robson (eds), A Century of Municipal Progress, 1835-1935, Allen & Unwin, London, 1935, pp183-97.

[23] Morrison, 'Highways and Transportation', p.194.

[24] William Plowden, The Motor Car and Politics in England, Faber & Faber, London, 1969, p.95.

[25] Morrison, 'Highways and Transportation, p.196.

[26] The IHE was formed in part, to improve the dialogue between the road construction and related industries and the engineers in the public sector. Though, since its inception it has had only one President from the private sector. Interview with Secretary, IHE. Note, too, that the British Roads Federation was founded in 1932, Hamer, Wheels Within Wheels, Appendix 1.

[27] James Drake, Motorways, Faber & Faber, London, 1969, p.36.

[28] Ibid.

[29] For example: 'It was satisfying to note that the motorway plan subsequently formulated by the Ministry after the War closely resembled the Society's (1938) recommendations'. Bill Harrison, 'Surveyors', p.45. At this time Colonel Harrison was Secretary of the Society. Other county surveyors have expressed the same view during personal interviews.

[30] Drake, Motorways, p.39.

[31] Ibid.

[32] Ibid.

[33] J B F Earle, Black Top, The History of the British Flexible Road Industry, Basil Blackwell, London, 1974, p.236.
The FCMI merged with the Asphalt Road Federation to become the Asphalt and Coated Macadam Association in 1967. The ACMA has been very active in arranging conferences and seminars for local government engineers on new materials and techniques of road repair and construction. It also has joint consultative committees with the DoE, the TRRL, and the County Surveyors Society. Since 1971 both its Director Generals have been former county surveyors.

[34] Rees Jefferys, The King's Highway, The Batchworth Press, London, 1949, p.130.

[35] Department of Science and Industrial Research, the Road Research Laboratory, German Motorways, 1946, HMSO, 1948.
It is worth noting that safety improvements became one of the prime benefits from motorways, stressed by the Institution of Highway Engineers.

[36] The Special Roads Act, 1949.

[37] William Plowden, The Motor Car and Politics, p.327.

[38] Drake himself comments, 'In 1945 when I was appointed Surveyor and Bridgemaster to my native county of Lancashire, I was excited by the thought that, at an early date, I might be designing a motorway to replace the trunk road north-south through my county. I set about furthering this scheme as soon as I could'. Drake, Motorways.
In 1951 he had prepared a model of a motorway junction for the British Roads Federation's stand at the International Motor Exhibition. Hamer, Wheels Within Wheels, p.32.

[39] This was the result of inadequate drainage, the first of many problems to plague the early motorways. Earle, Black Top, p.246.

[40] Earle, Black Top, p.246.

[41] Drake, Motorways, p.43.

[42] Earle, Black Top, p.23.

[43] IHE included almost all the country surveyors (in 1979 only about three remained non-members), central government engineers and private sector engineers connected with the construction industry.

[44] S E Finer 'Transport Interests and the Roads Lobby', Political Quarterly 29, 1958, pp.47-58.

[45] Cf. 'In 1964, Mr Ray Gunter, for Labour, declared: "I hope we shall look into the future and ensure that those who own private cars - everybody has the right to a private car if he wants it - are properly and adequately dealt with on the roads". Two years later Mr Peter Walker, for the Conservatives, said, "You must see the car as an important social development, bringing much pleasure to many people, so you must pursue an active policy to cope with it, not a negative attitude".' William Plowden, The Motor Car and Politics, p.385.

[46] Martin Painter, 'Whitehall and Roads: A Case Study of Sectoral Politics', Policy and Politics 8, 2 April 1980, p.167.

[47] Seventh Report of the House of Commons Estimates Committee, 1961-62 (HC 277).

[48] Their remit was 'to recommend what changes, if any, are required in the organisation and recruitment of the Ministry's staff in order to ensure the most effective implementation of the approved roads programme'.

[49] D.E.Reagan, 'The Expert and the Administrator', Public Administration 44, 3, Summer 1966, pp.156-7.

[50] Ibid, p.163.

[51] Ministry of Transport, Roads in England 1966-67, HMSO, 1967, para.3]3.

[52] This view was later expressed in their evidence to a Select Committee. See the House of Commons, First Report from the Transport Committee on the Roads Programme, 1980-81, HMSO, November 1980.

[53] Interview with a former county surveyor who held office in the CCS at the time.

[54] This view, that decisions tended to be taken at unnecessarily high levels within the Ministry, has been a perennial complaint of engineers at county level. In the course of interviews during this study there was virtual unanimity among engineers in the counties that this was so and that it was difficult to discover who was responsible for particular decisions.

[55] NEDO, Efficiency in Road Construction, HMSO, 1966; and the Second Report on Engineering and Road Construction, HMSO, 1967.

[56] Ministry of Transport, Roads in England 1966-67, HMSO, 1967.

[57] Roads in England 1966-67, para.3.26.

[58] Roads in England 1966-67, para.3.4.

[59] See note 53.

[60] Reported remark from a respondent.

[61] Roads in England 1966-67, para.3.7.

[62] North West - Lancashire, Cheshire
North East - Durham, West Riding
Midlands - Derbyshire, Staffordshire, Warwickshire
South West - Devon, Gloucestershire, Somerset
South East - Hampshire, Kent, Surrey
Eastern - Bedfordshire, Buckinghamshire, Essex, Hertfordshire
Roads in England 1966-67, para.3.5.

[63] Those schemes costing up to £1.5m in 1977, DTp, 'Guide to the Department of Transport', DTp, October 1977. In 1974 the DREs became Regional Controllers (Roads and Transportation).

[64] DTp, 'Report on the Study of Road Construction Units in Consultation with Sir Derek Rayner for the Minister of Transport'. February 1980, para.2.10.

[65] DTp, 'Guide to the Department of Transport'.

[66] DTp, 'Report of the Study', para.2.7.

[67] Ibid.

[68] Painter, 'Whitehall and Roads', p.177.

[69] Ibid, p.178.

[70] Under the Highways Act, 1959.

[71] These factors seemed to have contributed to delays in the construction of the M4. See Roy Gregory, 'The Minister's Line: or the M4 comes to Berkshire', Public Administration 45 (Part 1), Summer 1967, pp.113-128; (Part 2), Autumn 1976, pp.269-286.

[72] Painter, 'Whitehall and Roads'. Many highway engineers regarded motorways as a form of 'sculpture', James Drake, Motorways, p.209.

[73] Ministry of Housing and Local Government, People and Planning, Report of the Committee on Public Participation, HMSO, 1969.

[74] Under the Town and Country Planning Act, 1968.

[75] Changes were made in 1971 in the form of amendments to the Highways Act 1959, though these changes also gave the Minister

more discretionary powers, for example, to reduce periods allowed for objection. Further changes were announced in 1973 when the Department committed itself to the earlier publication of alternative proposals against the opposition of the roads men, who feared further delays. DoE, 'Participation in Planning' March 1973.

[76] Fred Mulley, the Minister of Transport after 1974, quoted in DoE, 'Relations between the Road Construction Units and the Public', Report of the Working Party, DoE, 1975, p.6.

[77] DoE, Heads of Division, Circular 37/74, July 1974.

[78] DoE, 'Relations between the RCUs and the Public', Appendix G.

[79] Ibid, Appendix G.

[80] The report expressed alarm at the way in which some inquiries have acquired the 'air of a TV court room drama because of the tactics of some organised opposition groups, and the presence and action of counsel'. DoE, 'Relations between the RCUs', p.9.

[81] Ibid, p.6.

[82] 'The Department's chief witness at the inquiry should be a well qualified engineer who is fully acquainted with the thinking behind the scheme and able to present the Department's evidence impartially. He must not appear committed to the proposal' (my emphasis). Ibid, p.10.

[83] 'Possibly the main worry of those objectors who are not motivated by a vested interest is the fear of change. The fear is real, it is often disguised in a comparatively trivial factual objection and remains even when the outward expressions have been shown to be groundless. These fears can only be allayed and the objector reassured by the objective presentation of facts, patient explanation of the proposals, the sympathetic treatment of the tangible grounds for objection, an explanation of remedies available, and the summary rejection of those people who take on these real fears to further their own vested interest.' Ibid, p.25.

[84] P.H.Levin, 'Highway Inquiries: A Study in Government Responsiveness', Public Administration,45, 1979, p.27; and Painter, 'Whitehall and Roads', p.179.

[85] The inquiry was not re-convened until almost four years later, in January 1980. The scheme was subsequently abandoned by the government following the Inspector's recommendation that improvements to the existing roads would suffice. For the protesters' view, see John Tyme, Motorways Versus Democracy, MacMillan, London, 1978.

[86] Terence Morris, 'The Highway Men', New Society, 9 September 1976, pp.549-50.

[87] DTp, Report of the Advisory Committee on Trunk Road Assessment, HMSO, October 1977.

[88] DTp/DoE, Report on the Review of Highway Inquiry Procedures, HMSO, April 1978 (Cmnd 7133).

[89] Painter, 'Whitehall and Roads', p.182.

[90] Secretary of State for Transport, Policy for Roads in England, HMSO, June 1978 (Cmnd 7132).

[91] NEDO Civil Engineering Committee, Design and Export, NEDO, HMSO, 1977. See also NEDO Structural Analysis Committee, How Flexible Is Construction? A Study of Resources and Participants in the Construction Process, HMSO, 1977.

[92] Minutes of the meeting of the National Annual Trunk Road Consultative Committee, 16 March 1977.

[93] British Road Federation, Basic Road Statistics 1980, p.31.

[94] DTp, 'Future Organisation for Road Construction: A Discussion Paper', July 1978.

[95] Ibid, para.16.

[96] These three points were made in ACC, 'Memoranda of Comment by the Association on the Government's Study of Road Construction Units', 1979. Although this was a response to the later Conservative Government's proposals, it has become clear in interviews that these points were made during the earlier negotiations with the Labour Government.

[97] Despite its title, APSPE does not include central government engineers.

[98] DTp, 'Future Organisation for Road Construction: Possible Options for Providing Staff to meet the Department's Supervisory and Procedural Role: A Discussion Paper', November 1978.

[99] Ibid, para.16.

[100] Comments by the staff side, reproduced in ibid, Appendix B.

[101] This emerged in interviews with county surveyors. Agreement had already been achieved in the Eastern Region but had not been formalised by the Minister.

[102] He had been appointed by the Prime Minister, Margaret Thatcher, and given a roving brief to examine possible waste across most government departments. His appointment was part-time and unpaid, otherwise he was a director of Marks and Spencer. He had previously worked in a similar role under Edward Heath in the 1970-74 Conservative Government. He had a small group of civil servants seconded to this team, who actually did the investigations, while he usually added the recommendations.

[103] DTp, 'Report of the Study of Road Construction Units in Consultation with Sir Derek Rayner for the Minister of Transport', February 1980, para.9.14.

[104] Ibid, para.9.20.

[105] 'The Department's lack of locus under the contract may make it harder to exercise financial control as a client'. Ibid, para.9.9.

[106] 'There are doubts about the performance of agent authorities which are largely to do with the terms of the agency agreement. The Department has no direct control of the members or quality of staff to be deployed on schemes nor, more importantly, of the progress of schemes; where the agent takes a different view from the Department on priorities, the ability of the Department to control its own programme is limited'. Ibid, para.9.8.

[107] This is how many country surveyors believe the DTp engineers saw privatisation. Given their close relations with the civil servants, it seems likely to be true.

[108] DTp, 'The Phasing out of the RCU Sub-Units. Ground Rules for the Action Study', 1980.

[109] Report of the delegation given to the ACC Transportation Committee, ACC, 'Report of the Planing and Transportation Committee', 10 July 1980. The 1977 NEDO Report had, of course, cast doubt on this assumption. Moreover, the ACC Transportation Committee Chairman publicly registered his view that 'many counties will be able to prove that they provide exceptional value for money'. Quoted in Municipal Journal, 14 March 1980.

[110] Reported in Municipal Engineering, 15 July 1980, 'Icy Response from County Surveyors to Phase-out of RCU Sub-units'. The 'four page critique' was signed by Colonel Leech, the then immediate past president of the CSS.

[111] Ibid.

[112] Ibid.

[113] These points are based on a personal interview with a leading ACC elected member.

[114] Policy for Roads: England 1980, HMSO, June 1980 (Cmnd 7908).

[115] First Report of the Transport Committee 1980-81, The Roads Programme, Vol II, HMSO, 25 November 1980, Minutes of Evidence for 16 July 1980, pp.35-39.

[116] Ibid, Vol II, pp.50-52.

[117] Ibid, Vol I, para.15.

[118] Under Maurice Milne, a former Deputy Director General (Highways), who sat with three advisers - a union representative, a representative of the Association of Consulting Engineers, and a

County Surveyor (the Secretary of the CSS), DTp, 'Road Construction Units: Clearing House for Staff', Memorandum from the Minister of Transport, October 1980.

[119] DTp, 'Road Construction Units: Action Plan to Phase Out the Sub-Units, Memorandum from the Minister of Transport', October 1980.

[120] Reported in Municipal Engineering, 23 October 1980.

[121] Reported in Municipal Engineering, 16 December 1980.

[122] Ibid.

[123] Ibid.

[124] The working party was chaired by a chief executive and was composed of three county surveyors, two treasurers and one ACC officer.

[125] ACC, 'The Future and Financing of the Trunk Road System', July 1981, para.1.3.

[126] Ibid.

[127] Ibid, para.10.2.

[128] Municipal Engineering, 28 July 1981. In the same issue Colonel Leech was quoted as saying: 'The Government has no business supervising the cutting of grass and prosecuting illegal signs on roadside verges. These are county jobs'.

[129] This report and its contents were mentioned in Municipal Engineering, 28 July 1981.

[130] Strictly speaking, in a letter to the associations on 23 April, just before SCHM met. See Chapter seven for details on SCHM, though it should be noted here that it is an officer level committee.

[131] See G D Spearing, 'Road Maintenance - The National Scene', in The Highway Engineer, March 1979, pp37-42. Spearing, an engineer, was Regional Director (Eastern Region) in DoE/DTp.

[132] Personal interview with a county surveyor.

[133] Personal interview with a civil servant. Also a Departmental Note on a meeting held on 19 June at the Department of Transport.

[134] Reported in Municipal Engineering, 11 August 1981. The consultants reported in February 1982.

[135] This point was made by a county surveyor at the CSS press conference to launch their Report, 'Transport in the 80s', 8 October 1981.

[136] See, for example, report of joint Institution of Civil Engineers/Institution of Highway Engineers Conference on Motorway Maintenance, Municipal Engineering, 7 October 1980.

[137] See Policy for Roads: England 1980.

[138] See report of contribution to ICE/IHE Conference by the DTp Head of Maintenance in Municipal Engineering, 7 October 1980. He maintained that the motorways would crack up faster than the projected rate of repair and maintenance.

[139] Painter, 'Whitehall and Roads', p.172.

[140] Personal interview with a county surveyor.

[141] Personal interview with a county surveyor who held office in the CSS at that time.

[142] The first was 'Transportation - The Next Twenty Five Years; an appraisal of the prospects and possibilities for the development of the country's transportation system up to the end of the century', CSS, 1978.
'The County Surveyors Society has, over a period of more than 90 years, sought to provide advice and guidance to assist the public authorities which its members serve in the determination of their policies for the discharge of their responsibilities, first as highway authorities and now as comprehensive transportation authorities. The Society feels that it has a responsibility, perhaps even a duty, as a body of experienced practitioners to state its collective views on the contribution which a sound, well-developed transportation policy can make to the economic well-being and prosperity of the nation, and to make an appraisal of the prospects and possibilities which are likely to affect the development of the country's transportation system', para.1.2.
The second report was 'Transport in the 80s', CSS, 1981.

[143] For example, the President of the Institution of Municipal Engineers at their 1981 Conference referred to local government reorganisation as a 'cruel blow to the profession' and said: 'One of the largest city authorities in this country now has virtually no municipal engineering input at the most senior management level and this at a time when one of the biggest tasks it is facing is urban renewal on a major scale'. Quoted in Municipal Engineering, 16 June 1981.

[144] At a rough count about 75% of 'directors of technical services' were engineers as opposed to architects, surveyors or planners. Source: Municipal Yearbook 1980.
Interestingly too, the non-metropolitan district engineers association is called the Association of Chief Technical Officers (ACTO). The Association was founded in 1977, partly at the instigation of the ADC, and one of its main functions is to advise the ADC.

5 The development of housing management

INTRODUCTION

In this chapter the development of housing management will be considered. Like municipal engineering its origins can be traced back to the nineteenth century, but while engineering became well established in local government during that century, housing management has only recently attained a comparable standing within local authorities. Much of this chapter, then, will be concerned with the obstacles to the professionalisation of housing management. As with municipal engineering, agencies of central government will be seen as playing an important role in the face of local indifference or hostility to professionalisation, while another important theme will be the significance of organisational factors in the development of the profession. In general, the case for this development has centred on considerations relating to organisational requirements rather than on considerations more typical of a public service profession. Although, as has already been noted, there has been a recent trend for local government professions to become increasingly concerned with organisational issues as local government bureaucracies have expanded.

THE NINETEENTH CENTURY ORIGINS OF HOUSING MANAGEMENT

During the middle nineteenth century some social reformers, appalled at the housing conditions in the towns and cities, began setting up housing associations. These associations were intended to provide sanitary housing for the working classes at a reasonable price, while at the same time giving the capitalist investor a good return on his capital - a combination of welfare and business motive called five percent philanthropy. These associations did not just provide housing but further concerned themselves with preventing the neglect and abuse of the buildings by tenants and with tenant welfare, especially children's welfare. A major reason behind the setting up of the associations had been the desire to reduce the appallingly high rate of infant mortality among the poor. Thus the provision of housing by the philanthropists gave them an excellent reason to intervene in the households of those workers whom they had rehoused. The object of these interventions was to instruct tenants in the methods of hygiene and household management, including the not unimportant matter of financial management in paying the rent. As it was the woman of the house who had to be instructed in the conduct of her household, this welfare role as defined as women's work. Although, inevitably, the housing associations were funded and run by men, it was not considered appropriate for men to have such close contact with the opposite sex and, besides, welfare was seen then as a peculiarly female preserve.[1] Accordingly the associations set about recruiting women from their own class to perform this work. At first, as was considered appropriate for ladies, the work was done on a voluntary basis: only later was it recognised as warranting remuneration.

The existence of these associations and the acceptance of the importance of the female role within them gave some women the opportunity to build careers. Housing management was one of the few areas in which women could find worthwhile employment, which would eventually become paid. The first woman to exploit this opportunity and encourage others to follow her example was Octavia Hill. She is traditionally regarded as the founder of housing management and established, what has become known as, the Octavia Hill System of housing management. The System stressed the contribution that good housing management could make to the upkeep of the buildings and the welfare of the tenant. It was a very labour intensive approach, at least by today's standards, as it rested upon managers monitoring tenants during regular visits to collect the rent. So this point of intervention between the tenant and the manager was not simply an economic transaction but also provided an opportunity for the manager to check on the tenants' welfare and for the tenant to raise any problems such as repairs. From the first, then, in the voluntary sector which adopted this system, housing management was conducted in an integrated way with housing managers responsible for rent collection, allocations, resolving tenants' disputes and the maintenance of the buildings. In order to cope competently with these tasks the managers required training and it was this requirement that Octavia Hill set out to fulfill. The training was along apprentice lines with the novice housing manager being guided by the more experienced. There was of course no formal qualification at this stage, but there was an expectation in many housing associations that housing managers should have undergone a period of training with Octavia Hill or one of her followers before being allowed major responsibilities.[2]

During Octavia Hill's lifetime no housing management association was created. It was not until four years after her death that an association was formed. Significantly the association was formed in 1916 during the First World War when women had been called in to the Ministry of Munitions to manage new housing built to accommodate munitions workers. The women managers gathered to discuss the new challenge raised by the War and decided to form the Association of Women Housing Managers. Unfortunately, the advances made during the War by the women were lost when the men returned and most of the women felt morally obliged to surrender their new jobs to the returning soldiers.[3] But during the 1920s the re-named Society of Women Housing Estate Managers began to recruit larger numbers of women housing manager's employed to manage some of the new municipal housing built after the war.[4]. In 1933 the Society introduced the first formal qualification in housing management, this qualification was examined by the Royal Institute of Chartered Surveyors for the Society, an arrangement that lasted until the 1960s.

THE BEGINNINGS OF LOCAL AUTHORITY HOUSING MANAGEMENT

Local authority house building had begun as early as the middle of the nineteenth century but it was not until after the Housing of the Working Classes Act 1890 that public sector building began on any scale. Even then only 24,000 units were constructed before the First World War.[5] It took the serious post-war housing shortage for the government to commit themselves to expand the numbers of houses built by the municipalities which they did by providing an exchequer subsidy; provision for this new building programme was made in the Addison Act of 1919. There were a number of further legislative changes during the 1920s and 1930s that need not be detailed here - sufficient to make the point that the building of public housing continued at a high level until the beginning of the Second World War.

Before the 1919 Act many of those authorities which had built houses allowed them to be managed by private sector estate agents or solicitors. However, with increased building this type of arrangement was seen as unsuitable and the Ministry discouraged it.[6] But the arrangements that authorities made to manage their housing stocks themselves differed widely and as yet there was no central guidance. At this time no specific housing departments were established, the different tasks involved in housing management were divided up among the existing departments - rents were collected by the treasurer's department, housing allocations made by the clerk's department and maintenance carried out by one of the technical services departments. There were, however, some attempts to apply the Octavia Hill System of integrated housing management to municipal housing, interestingly this form of intensive housing management was usually adopted on those estates seen as having management problems.[7]

Inevitably the great majority of those involved in the management of the new municipal housing at this time were male local government officers. The option of joining the Women's Society was not open to them, not just because of their gender but also because of their lack of training. However some municipal housing managers began meeting together in the Midlands initially to discuss common problems and pool experience.[8] It was out of these meetings that the idea of a national association emerged. In 1931 the Institute of Municipal Estate Managers was formed, the forerunner of the present Institute of Housing after three further changes of name. This new Institute was predominantly but not exclusively male in its membership.[9] Unlike its sister association the Institute did not set an examination qualification for entry. Instead entry was restricted to those officers who had reached a senior position in their authorities, while their more junior colleagues were excluded from membership.[10] Thus membership was open to any officers belonging to those professions then associated with housing management. So the Institute of Housing began life as a co-ordinating assocation, not as a qualifying body - it did not acquire a method of entry by examination until 1946.

The new Institute made slow progress in its early life. By 1935 only 17% of all local authorities had appointed housing managers, while in most areas the treasurers, town clerks, engineers or surveyors took responsibility for managing the housing stock.[11] In 1939 the Institute had a membership of 261 of whom 78 were chief housing officers with the rest in more junior posts.[12]

The other important development for housing management during this period was the creation of the Central Housing Advisory Committee which was set up under the Housing Act 1936 to advise the government on housing matters. One of its first actions was to establish a Housing Management and Housing Associations Sub-Committee which immediately began the first central government investigation into housing management practice within local authorities. The Sub-Committee's remit was to consider, inter alia, the general question of the management of housing estates by local authorities with special reference to the employment of trained housing managers.[13]

The Sub-Committee went about its work by conducting an investigation of the state of housing management as it then was. Its members investigated eight housing authorities and were appalled at the conditions that they found on many estates. Their major concern was the failure of these authorities to follow up their tenants after they had been re-housed from slum clearance schemes. They found cases, for example, where the tenants had no furniture and where repairs had not been completed. The assumptions of the Sub-Committee were paternalistic, and they felt that authorities were failing in their duty, the municipal landlord was considered to hold a position of special responsibility and privilege, responsibility because he was the custodian and trustee of property provided through the combined resources of the taxpayer and the ratepayer, privilege because he must be the counsellor and friend of an increasing number of tenants coming under his charge.[14]

As well as being a call to authorities to shoulder their responsibilities, it was also a succinct statement of the conflicting roles of local authority landlords as both property owner and as welfare agency. The report went on to define the central tasks of housing management as the application of skill in treating the tenants in order to help them to preserve the value of the housing both for themselves and for the community. Thus housing management was seen as a form of social education and aimed at teaching a new and inexperienced community to be housing minded.[15] The views of the Sub-Committee were strongly influenced by the evidence detailing the Octavia Hill System submitted to them by the Society of Women Housing Managers; they had too seen the System in practice during their visit to Rotherham where it had been applied on one estate. However, they did not confront the major drawback of the System, a drawback that the Institute had mentioned in their evidence, which was its labour intensive nature.[16] In their own evidence the Society stated that it required about one manager for each 300 tenants.[17] The Institute, more realistically, maintained that authorities with large housing stocks could not afford such intensive management practices. Their evidence too cast doubt on the importance of the social welfare functions of management that their sister association had advocated. In the Institute's view this function should have been separated from other management functions such as repairs and rent collection. Undoubtedly this viewpoint must have been shaped by the fact that the membership of the Institute was drawn from those professional groups then responsible for housing, professions which could have had little interest in creating a new occupational group with its own set of distinct and inter-related tasks. Rather their interests laid in maintaining the fragmentation of these tasks. In contrast, the Women's Society had already developed a professional account that presented these tasks as interrelated and as centred on the normative task of social welfare. The Institute's view was that of people who were first and foremost local government officers, whereas the women housing managers were first and foremost committed to social welfare objectives.

The Sub-Committee concluded in favour of the type of integrated approach advocated by the Women's Society. The Report recommended that a range of tasks should be recognised as related and as part of housing management's responsibility, these included allocation, the after-care of tenants, the ordering of repairs, and the provision of community facilities. In order to cope with this range of tasks it further recommended that housing managers should be trained. A proposed syllabus for such training was outlined in the Report, and it included a grounding in both welfare and in more technical skills. The managers were to be trained to give advice on the availability of unemployment and other benefits, nutritional levels, the national savings scheme and activities for local community centres.[18] On the more technical side, training was to encompass subjects such as the elements of economics, housing law, the basics of building and sanitation, and basic bookkeeping.

This Report was distributed to local authorities and the contents recommended to them in an accompanying circular. It represented an important milestone in the emergence of a new profession in the housing field. New tasks, which had not been recognised in the majority of authorities, had been recognised in an official central government report. But these tasks and the relation between them and existing managerial tasks had already been widely recognised in the voluntary housing association sector and CHAC was now recommending that local authorities should learn from this example. Significantly the pressure for change had not come from the occupational association based in local government but from the Society whose members worked predominantly outside local government. Moreover, the stimulus needed to form the basis of a new professional grouping in local authorities had come from a body based in central government. However, the intervention of the Second World War halted any further change in response to the Report.

HOUSING MANAGEMENT: THE POST-WAR PERIOD UNTIL 1960

The Second World War had a profound effect on all social services. In its aftermath spending on these services increased and the government took on responsibility for aspects of the individual's life that it had never had before. Education was made compulsory up to the age of sixteen; unemployment and national assistance benefits, free medical care and a state pension scheme were introduced among other new welfare measures. At the same time the Government took a greater interest in the provision of public housing. There was, after the War, a serious shortage of accommodation as the result of bombing and the post-war marriage boom as servicemen returned. The Labour Government's response to this shortage was to begin an ambitious programme of house building in the public sector, a policy which in the beginning had bipartisan support. Although later on the two main political parties differed over the relative emphases they placed on public and private sector new building, they did not differ in the importance they placed on increasing the stock in absolute terms. In fact post-war housing policy was dominated by competition between the two parties over which could build the largest number of houses. This competition effectively placed problems of the management of the housing stock in the shade until the late 1960s. There was also another important housing issue that received almost as much prominence as new building, this was rent control in the privately rented sector, an issue over which the two main parties were divided.

The immediate post-war house building boom took place under Aneurin Bevan, the Minister of Health at the time. It was Bevan who personally insisted that public housing should continue to be built by the local authorities rather than by a new central government agency of the type favoured by many of his colleagues in the Labour Party.[19] Bevan felt strongly too that the housing provided by the state should not be in any way sub-standard. So his Ministry required local authorities to build to certain defined standards.[20] At the same time the Ministry appointed the first permanent housing management advisor, Molly Empson, in 1946.[21] It was, perhaps, significant that the Ministry appointed a member of the women's Society and not a member of the local authority based Institute.

In the year before Molly Empson's appointment, CHAC published a further Report on the management of municipal housing estates.[22] This second Report was really just a supplement to the 1939 Report but it did reflect a small change in attitude since that Report. Local authorities were facing a different problem in relation to allocation to that which they had faced before the War. Then allocation was, for the most, part unproblematical insofar as those excluded from public housing were not politically organised or vociferous. But the War had created at least two new groups of people who felt that they had a prior claim on available housing, returning servicemen and their families and those whose homes had been destroyed by enemy action. For the first time housing authorities found themselves dealing with applicants who felt that they had a right to housing and were willing to protest when that right was denied them, while at the same time the councils were facing a shortage of housing. In these circumstances CHAC issued this Report to give councils advice on how to cope with the new problem, exhorting them to adopt equable and flexible selection schemes in response.

CHAC was unhappy with the existing selection practices, as many local authorities had no formal selection schemes at all. The Committee thought that the adoption of some ready and uniform method of classification of housing applicants was imperative.[23] In addition, they recommended that these schemes should be publicly available and that elected members in committee should not be told the names of prospective tenants. These three changes would, the Committee thought, allay suspicions of favouritism among applicants.[24] It is worth noting that it was not until the 1980 Housing Act that the publication of selection schemes became widespread. Otherwise CHAC recommended that councils should adopt a very broad view of the classes of people that they should consider in letting their housing, but should give priority to need rather than any other considerations.[25] This meant that they should not seek to apply rigid residential qualifications which could cause hardship given the amount of war-time movement amongst the civilian population. Interestingly both the Institute of Housing and the Society of Women

Housing Managers in their evidence to the Committee expressed their opposition to residential qualifications.[26] Nevertheless, as will be seen in the rest of this study, residential qualifications and restrictions have persisted up to the present in the great majority of authorities.

In accordance with the recommendations of CHAC, the government legislated to remove the earlier obligation on local authorities to provide housing for the working classes only. This action implied that public housing should not be seen as just accommodating particular deprived groups and that a stigma did not attach to those people who lived in such housing. In the same year further advice was given to local councils on the selection of tenants and transfers and exchanges.[27] This Report elaborated on the advice and exhortations of the previous Report. It gave some more detailed advice on the formulation of selection schemes. But it stressed too, more so than the earlier Report, the importance of mobility both in personal terms and in terms of economic growth. Thus transfers and exchanges were taken as ways of assisting worker mobility, and this time CHAC condemned residential requirements as unnecessarily hindering mobility. This issue of mobility was to become a recurrent one during the post-war period, with central government pressing for change, the local authorities resisting, but the government refusing to legislate on the matter despite the continuing failure of local authorities to implement the necessary changes.[28]

The post-war period brought significant changes in the organisation of the professional bodies. In 1938 the Institute established entry by examination for the first time but retained the earlier practice of admitting members on the basis of their occupancy of senior positions in authorities. Unlike the Society, the institute administered and examined the qualification themselves. In the same year the Royal Institution of Chartered Surveyors introduced a housing management section of their professional examination and a housing managers' certificate designed for those who did not aim at a full professional qualification.[29] Thus many of the older housing managers hold a surveyors' qualification. In that year the Institute had 666 members, including 206 students.[30] During the late 1940s the membership of the Institute rose steeply and by the end of the decade there were over 900 full members.[31] But after this surge in membership, the numbers remained more or less constant during the 1950s and 1960s. Similarly the Society membership changed little over the same period until their amalgamation with the Institute in 1964. The post-war period brought significant changes in the organisation of the professional bodies.

During the 1950s the building of public housing continued on a large scale. This new housing was increasingly built in the form of large housing estates on green field sites on the edge of towns. The reasons for building in this way were that it was economic in terms of finance and effort. Obvious economies of scale can be gained from building large numbers of units in the same location; and from the point of view of administrative convenience it is easier to arrange for development on one site rather than on several dispersed sites. The pressures from the centre to build houses and the subsidy systems instituted by the centre reinforced these more local considerations. Another important feature of the subsidy system during this period was the encouragement it gave to the building of high density housing and in particular high-rise housing. In 1955 Duncan Sandys, the then Minister of Housing and Local Government, announced that the system would be changed to offset the extra cost to authorities of building blocks of flats above five stories. This had the effect of encouraging authorities to build high-rise blocks.[32] The Government's reason for this change was to increase the level of house building which was still a hot political issue between the two parties.

The next housing management report produced by CHAC in 1959 was concerned with the management problems raised by the new, large housing estates. In contrast to the 1938 Report, this Report emphasised the financial side of management rather than the welfare side. The keynote statement of the Report should be contrasted with the statement quoted earlier from the 1938 Report:

> We believe it necessary at the outset to emphasise that local authority houses and flats represent a considerable part of the nation's wealth, and that it is the responsibility of local authorities to treat their management as an important business enterprise which must be conducted with due attention to the financial, as well as to personal aspects. The houses must be well looked after, fairly rented, and economically managed. An adequate system of costing is as essential to good housing management as to any other business undertaking, so that the houses do not become a burden instead of an asset to the community.[33]

Clearly, in the intervening twenty years, the emphasis had shifted away from the earlier paternalism. Although the 1959 Report did not ignore the welfare aspects its case for unified housing management rested not on welfare arguments as in 1938 but on the need for sound businesslike management. Reading between the lines of the Report, it seems that the government was concerned that councils were failing to build and manage their large new estates economically and efficiently. In fact the Report recommended an inquiry into the costs of housing administration.[34]

In the Report CHAC was concerned that councils were failing to take account of maintenance and management problems in the design of new estates. It attributed this failing, at least in part, to the organisational separation of the design functions from the management functions, which was such that, from the outset of work on particular house or estate designs, sufficient consideration was not given to the implications for future maintenance or for tenant satisfaction. This situation could be remedied, CHAC proposed, by having a single committee responsible for building and management,[35] and by involving the housing manager at the design stage,[36] as the latter would have first hand experience of the structural problems of the housing once built and of the tenants' point of view.[37] The Committee did not recommend that design and management should necessarily be combined within a single department, just that they should be combined at committee level. But they did take the view that the various tasks that comprised housing management would usually be best performed through a single department.[38] The reasons for creating unified housing management departments were twofold. Firstly, given that the various activities involved in the care of estates were closely interdependent, only with full co-ordination under a single head could the best results be achieved, and overlapping, duplication or neglect of detail be avoided.[39] Secondly, the advantage of the single department was that the tenant dealt with only one department during his tenancy and his application for that tenancy, so it was simpler from the tenants' point of view.[40]

It is worth looking closely at the type of competences the Report regarded as necessary for the housing managers in these unified housing departments. CHAC thought that the housing manager should combine a wide variety of technical skills with administrative ability.[41] These were essential to cope with the wide range of housing management functions. As well as administrative ability, housing managers were expected to have some knowledge of housing and property law, housing finance and estate accounts, building construction and methods and materials used in maintenance, local government administration and an understanding of the social services provided by the community.[42] This list underlined the very diverse and discrete bodies of knowledge on which housing managers had to base their claims for separate housing management departments under their control. It was this very diversity that contributed to the difficulties they experienced in trying to establish a coherent professional identity.

In the survey of housing authorities undertaken in the course of the Committee's work, they detected a movement within local authorities towards unifying these management tasks within a single department. Thus half of the authorities included in the survey had a proper housing department, although eleven had no housing department at all. But already nearly all the housing managers dealt with applications and lettings, general estate supervision and housing welfare.[43]

The Report not only contained some recommendations for organisational change but also for improvements in the training of housing officers. CHAC saw the efficient management of the new large estates owned by local authorities as dependent on the use of trained and qualified staff. Yet at the same time CHAC recognised that there was a shortage of recruits to the profession. This shortage was attributed to the low status of housing management and the fact that the salaries were regarded as unattractive by potential entrants.[44] CHAC's recommendation was that this state of affairs should be remedied by authorities offering more training places to recruits and by improving the career prospects of housing officers.

HOUSING IN THE NINETEEN-SIXTIES

The production of new housing continued apace during the 1960s. In the last years of the Conservative Government the 1961/62 boom led to increased output, with the councils' share of new building actually increasing to its highest level since 1955.[45] This reflected the greater ease with which the Government could expand house building by local authorities than by the private sector.[46] The other important change initiated by the Conservative Government was the creation of the Housing Corporation in the 1964 Housing Act. The new Corporation was intended to channel public funds towards the housing associations, which were seen as widening consumer choice in housing, in particular for those groups which did not qualify for council housing but could not afford to buy their own house.[47] The creation of the Housing

Corporation laid the basis for the future expansion of the housing association sector under the later 1974 Housing Act. One of the major issues in the 1964 General Election was the competition by the main parties over promises for the future level of house building. The new Labour Government came into office committed to building half a million houses a year,[48] so surprisingly the first Government housing White Paper expressed the view that building for owner-occupation was the 'normal' state of affairs and that public sector building was only a short-term expedient to meet 'exceptional needs'.[49] This marked a break with the previous Labour approach to council housing. They had not regarded it as a form of residual provision for the poor alone. It also indicates the emergence of a bipartisan consensus on the importance of assisting the owner-occupier and those wanting to buy their own house.

One important aspect to both the Conservative and the incoming Labour Government's house building programmes was the encouragement that they gave to the building of high-rise flats. In 1964 the Government set up the National Building Agency and a special division in the Ministry of Local Government to promote the use of system building by local authorities in the construction of tower and slab blocks of flats. Undoubtedly the Government saw the use of such techniques as essential to the achievement of their housing targets. Under pressure from the Government, architects and industrial interests, many authorities adopted these new techniques.[50] The application of these new and untested building methods was to have serious consequences in the future.

In 1965 the Central Housing Advisory Committee decided to follow up the recommendations contained in their 1959 Report regarding the establishment of unified housing management departments. The Committee sent out a questionnaire to all housing authorities and 1135 replied. The results indicated some movement towards the development of the profession of housing management,[51] though the patterns of the housing management organisation still varied widely among local authorities. Since 1959, 118 authorities had appointed housing managers for the first time or had set up separate housing departments.[52] But still only just over half employed a housing manager, and fewer had their own department - 418 out of the 1135. The other topic, which the survey examined, was the progress that had been made in implementing CHAC's earlier recommendations on the training of housing officers. The Committee was concerned with the lack of progress in training that the survey revealed and, in particular, with the failure of some of the more important authorities to follow their recommendations.[53] As a result discussions began between the Ministry, and the local authority associations and between the Ministry and the Institute of Housing Management on ways of improving training arrangements.[54] Clearly progress towards the full professional recognition of housing management within local authorities was slow. It should be stressed that the push for enhanced professional status continued to come from central government rather than from the authorities themselves who, if anything, dragged their feet on the issue.

One very significant change, which took place during the 1960s, was the growth of pressure groups concerned with housing policy. There were already pressure groups in existence in housing but the new groups founded at this time were much more active in public campaigns than the older pressure groups which were, and are, much more part of the housing 'establishment' in their approach. Of these earlier groups, there are several representing the interests of property, like the House Builders Federation. Then there are several professional associations with members working in the housing field which express their views from time to time; for example, the Royal Association of British Architects and the Royal Institution of Chartered Surveyors (which has many estate agent members as well as local government members, including some housing officers). Finally, there are the pressure groups which do not, at least explicitly, have any allegiance to particular vested interests. From the point of view of this study, the principal pressure group in this category is the Housing Centre Trust, although there are two others the Town and Country Planning Association and the National Housing and Town Planning Council. The TCPA's interest in housing is quite specialised. It grew out of the Ebeneezer Howard garden city movement and favours the continued dispersal of population from the cities to the new towns and growth areas.[55] The NHTPC, unlike the other two, is an elected member association, formed just before the First World War in the wake of the expansion of municipal housing and planning powers. Its main work is the publication of a regular journal and an annual conference with a commercial exhibition, otherwise the local authority associations have taken over its role as a platform for municipal housing.[56] Thus the Housing Centre Trust is the main non-campaigning pressure group in the field. Its membership includes officers, elected members, academics and others interested in housing matters, for the most part, from a centre-left perspective. Nevertheless the Centre insists on its non-partisan character, though it has taken a stance on partisan issues such as the level of subsidy for different types of housing. It presents itself as a group of neutral and informed people. It works mainly through usual channels, its forte being well-researched papers and not vociferous public campaigning. So it has attracted some of the more intellectual housing officers interested in pressing for changes in housing policy. These officers have tended to use the Housing Centre as a substitute for the Institute of Housing which has been unwilling to take up stances on housing issues and lacks a tradition of producing well-researched policy papers. The Centre has the further attraction of including some elected members whom many officers regard as the key people to influence. As well as producing policy papers it also publishes a bi-monthly journal, 'Housing Review', and holds frequent seminars for both officers and members on housing matters. It has probably been responsible for assembling more housing officers together than any other body.[57]

So the new housing pressure groups, which emerged in the 1960s, were different both in approach and in membership from these older pressure groups. The background to their emergence was the election of a Labour Government in 1964 and its perceived failure to solve the problem of homelessness. This was in contrast to the optimism of earlier statements by Labour spokesmen during the Labour election campaigns of 1964 and 1966. This disillusionment with the Labour Government and with conventional parliamentary politics, perhaps as the result of excessive expectations of what could be achieved, contributed to a new interest in campaigning pressure groups. And concern with the problems of homelessness was beginning to be reflected in newspaper stories about the plight of the homeless.

The best known pressure group formed at this time was Shelter, the National Campaign for the Homeless. This was formed in 1966, just a month before the now famous BBC film 'Cathy Come Home' which vividly dramatised the problem of homelessness and provoked a tremendous public response.[58] Shelter had as its objects getting more money into public housing programmes, persuading local authorities to adopt more humane policies and pushing for more rehabilitation of the older stock.[59] Their approach was unusual in two ways. Firstly, from the beginning the group supported its own housing associations, and housing association people had been prominent among its founders. Moreover, it subsequently set up its own housing advice centre in 1970.[60] Secondly, Shelter intended not just to campaign at the national level but also at the local level. From early on in its life local groups were formed and inevitably came into conflict with their local authorities. Their approach was, and is, to take up individual cases and campaign upon them, using the particular case to dramatise the underlying problem, usually the failure of a local authority to respond to homelessness in a humane fashion in Shelter's view.[61] This approach has led to considerable hostility towards Shelter among many housing officers who feel that they and their councils have been unfairly pilloried.[62] Shelter itself has been critical of housing management's failure, in its view, to professionalise fully in the sense of accepting a moral responsibility for the tenants in public housing.[63]

The growth of such groups as Shelter reflected a further notable development in the housing field during the nineteen-sixties - a major shift in thinking about housing policy and about the role of the public sector. As has been seen, until the late sixties housing policy had been dominated by the political imperative of building more houses, while the public sector had been seen as providing almost exclusively for families that could not afford to buy their own housing. Thus the housing problem was seen in simple terms as a national level shortfall between housing need or demand and the supply of housing. Commentators came to question this straightforward but dominant assumption about the nature of the housing problem. This questioning reflected a changing debate about the nature of deprivation and, what has been called, the rediscovery of poverty in Britain. It was this rediscovery which was closely related to the emergence of the new pressure groups. Poverty and deprivation were seen as being located within inner city areas by-passed by the growth of affluence in the rest of society. So the new approach to social policy, heavily influenced by the American example in both practice and diagnosis of social problems, tended to stress the locational aspects of deprivation.[64]

Some academics, like David Donnison and Barry Cullingworth, pioneered a new definition of the housing problem. They argued that far from there being a single national housing problem, the national housing system should be regarded as a series of fragmented, local housing systems each with their own particular combination of problems.[65] Accordingly, they argued that policies relating to housing were best co-ordinated at the local level. Thus, policies relating to the new towns, council housing, the privately rented sector, house purchase and slum clearance should be considered as inter-related at the local as well as at the national level and not treated separately as had been the case until then.

This new policy emphasis began to emerge in the changing housing policies of the Labour Government. Interestingly, the Labour Government's interest in the idea of local authorities playing a wider role in local housing reflected, in part, an impatience with those Conservative authorities which refused to build and improve housing at the rate that the Government wished to see. Accordingly, to speed things up at the local level, the Government placed new housing obligations on authorities. In the Housing Act 1969 local authorities were urged to be more interventionist. They were placed under a duty to assess what needed to be done about unsatisfactory areas within their boundaries, empowered to declare general improvement areas, to acquire land and buildings in these areas for the improvement of houses, the environment, or for clearance.[66] It should be added too that these doubts about the effectiveness of the existing councils as housing authorities contributed towards the pressure for local government reform.[67]

THE CULLINGWORTH REPORT

The next CHAC Report on housing management should be seen against the background of these changing views on housing policy. It was chaired by one of the leading exponents of the new approach, Barry Cullingworth the then Professor of Urban and Regional Studies at Birmingham University. Generally recognised as the most important central government report ever produced on housing management, the Report raised most of the issues that were to concern those involved in housing during the nineteen seventies. The Committee's terms of reference were just to examine the allocation of council houses but, in the event, they interpreted these terms liberally. Their Report focussed on two major aspects of housing policy. Firstly, the implications of the decline in the privately rented sector were stressed. In particular, it was argued that many people who would formerly have been housed in the private sector would be obliged to look towards the public sector for accommodation. Local authorities would have to take on the responsibility for the housing needs of these people and to plan future housing provision in accordance with these needs. Secondly, the Committee noted the trend towards smaller households as reflected in the falling rates of room occupancy and was concerned that councils should begin to change their policies and provide a wider range of housing suitable for special groups, particularly the elderly, rather than continue providing the usual, standard three bedroom dwelling.

As a necessary step towards changing the policy direction of local authorities, the Committee recommended that local authorities' statutory responsibilities should involve both assessing the housing needs of a locality and actively searching for housing difficulty within the local community. In addition, the Report emphasised that the local housing system was more complex than had previously been assumed so that authorities had to have a clearer, deeper and more detailed understanding of the housing situation in their area:[68]

> ...we no longer have a single national housing problem: we have a large number of local housing problems of great variety. It is therefore essential that local policies be based upon a well-informed understanding of individual areas and the context in

> which they arise. Our first recommendation is, therefore, that local authorities should take steps to ensure that they are better informed of the housing situation in their areas.[69]

Significantly, the Committee was also critical of the allocation policies then operated by many local authorities. There were three main aspects to these policies: the selection of tenants, residential qualifications and tenant mobility. The Committee was unhappy with the involvement of elected members in the allocation of individual tenancies. In their view this practice opened up the possibility of patronage and at the very least meant that councils could be open to the charge of favouritism. Accordingly, the Report recommended that the contribution of the elected members should be restricted to fixing allocation policies, which would then be left to the officers to administer. As far as the actual content of the allocation policies was concerned, it was considered that more weight should be given to social need as against other factors, since the Committee believed that too many prospective tenants in bad conditions were not being allocated tenancies. In particular, the Report drew attention to special groups, such as the homeless, which had been neglected as a result of restrictive allocation practices. Related to this, the Report censored those authorities which adopted a moralistic attitude towards the poor, differentiating between the deserving and undeserving poor.

Secondly, the Report returned to the issue of residential requirements. The earlier CHAC reports had condemned these requirements because they conflicted with the allocation of tenancies on the basis of need. The Committee's view, then contrary to that of almost all housing authorities, was that no-one should be precluded from applying for, or being considered for, a council tenancy on any grounds whatever.[70] They even recommended that authorities should be placed under a statutory duty to maintain an open waiting list, the only recommendation in the Report that proposed further central intervention. Open waiting-lists were seen as an important part of persuading local authorities to take on a wider range of housing responsibilities.

Thirdly, the Report was concerned with the problems of mobility faced by council tenants and prospective tenants wanting to move into an area. Improvements in exchange and transfer policies, as well as the abolition of the residential requirement, would not just enhance the freedom of tenants but would also assist economic growth by removing an obstacle to worker mobility.

This Report was ambitious in terms of the expanded responsibilities that it proposed should be taken on by local authorities. It was enthusiastically taken up by some of the more ambitious housing officers, almost as a professional manifesto. While others dubbed it the 'Yellow Peril' (from the colour of its cover) and several asserted that they would make sure their committees would never see it.[71] The Seebohm Report on local authority social services had already, in the year, before, advanced the case for comprehensive housing services operated by local authorities, as part of their recommendations for the improved co-ordination of social services.[72] So the professional community found themselves presented with a new professional account that took them beyond what had formerly been the central concern of the profession - the landlord-tenant relationship within local authorities.

The comprehensive housing service was one step on from the earlier idea of unified housing departments. This new rationale had originated outside the profession, reflecting the concerns of the social welfare lobby and, as will be seen in the following section, reflecting the new managerial ideas spreading through local government at the time. As such it raised not only opportunities for the profession but also new challenges. The professional community's response to these opportunities and challenges will be discussed in the following sections.

LOCAL GOVERNMENT REORGANISATION AND THE NEW HOUSING DEPARTMENTS

The comprehensive housing service chimed in well with the ideas in good currency at that time. The notion of an enlarged housing department containing the major housing activities of an authority was in accord with the stress placed by the corporate management movement upon programme-based departments as against traditional profession-based departments. The major organisational blueprint for the new authorities, the Bains Report,[73] to which the Institute of Housing had submitted evidence, argued strongly for a committee structure that reflected programme areas rather than professional jurisdictions. And, although the Report did not generally favour the idea of the departmental structure being based solely on programme areas, in the area of housing the Report did recommend a shift in this direction. The recommendation was that 'a chief officer or director of housing should be responsible for the total housing function, including management, assessment of need, improvement, slum clearance, and any advisory service'.[74] As significant was the recommendation that directors of housing should sit on the management teams of the new authorities while such traditionally well established figures like the chief architect and chief engineer were not included on the management teams. These recommendations were seen by the profession as an important official recognition of their claims to a wider departmental remit, even though Bains said nothing about whether these departments should be headed by qualified housing officers.

Reorganisation, then, was seen by those leading housing officers on the Council of the Institute of Housing as a major opportunity to enhance the power and standing of the profession. It became a major priority among them to seize the initiative and to press further the new comprehensive housing service approach. In 1972 the Institute produced its own set of recommendations for the new housing departments along comprehensive lines.[75] As one member later commented, 'We were attempting to have our say over Reorganisation'.[76] The Report was intended to provide facts and arguments that 'officers could use as leverage on members back in their authorities',[77] during a period of organisational flux when organisational blueprints were being anxiously sought by new elected members faced with the colossal uncertainties incumbent on setting up new organisations. The Report was a brief document that gave details of what existing authorities had done, especially those in London, most of which had set up enlarged housing departments after London Government was reorganised in 1963/4.

The Institute was pleased with the results of Reorganisation. They conducted a survey of the housing departments in the new authorities which suggested that the profession had gained more ground than they had dared to hope:

> When the first discussions took place at the Institute Headquarters, my recollection is that members thought that if about half of the new Housing Departments were headed by a Chief Officer it would be very good, but we actually have 88 per cent of the new Departments established in this way. So instead of the toehold forecast by some, there is a substantial foothold and more.[78]

The survey revealed that 75% of the new chief housing officers were members of their authorities' management teams. While in the 54 authorities with populations over 150,000 51 had housing departments run by an officer with chief officer status, though in only 40 of these were the chief housing officers also members of the management team. Even so the results of the survey indicated that many of these new housing departments were not directly controlling all the functions related solely to the tasks of housing management. In a third of all authorities, rent collection was administered outside the housing department, though in 92% of the larger authorities (those with populations over 150,000), rent collection was a housing department function. In the smaller authorities (population below 150,000) only 59% (173 out of 292) of housing departments included rent collection.[79] A similar pattern was revealed by the survey on rent assessment; again in 78% (43 out of 54) of the larger authorities this function was performed by the housing department, while in only 50% (139 out of 290) was this the responsibility of the housing department. This pattern was repeated yet again over rent accounting which was done by only 46% (133/292) in contrast to 81% (44/54) of departments in the larger authorities.

This discussion indicates that in the smaller authorities, housing departments were very much less likely to be performing those functions related to finance. This suggests that in the smaller authorities, which are all non-metropolitan districts (there being no metropolitan districts of under 150,000 people), the treasurers or accountants had been reluctant to lose their traditional involvement in housing management. In contrast, in the larger authorities, which are predominantly metropolitan, fewer treasurers' departments were involved directly in the housing service. This reflects, in part, the fact that in the shire districts housing represents the biggest spender, whereas in the metropolitan districts it represents a much smaller proportion of the total budget as these authorities are also responsible for other big spending services, in particular education. To those within the treasurer's department in the metropolitan districts, therefore, the temptation to become involved in housing management issues was considerably less than it was the case of treasurers in non-metropolitan districts.

In the next year, after local government reorganisation in England and Wales, Scottish local government was reorganised. Scottish reorganisation brought two further relevant reports that referred to the housing service and gave support to the idea of strengthened housing departments. The first was the Patterson Report,[80] the Scottish equivalent of the Bains Report, containing recommendations for the departmental and committee structures of the new authorities, and which included the idea of enlarged housing departments. The other

Report, the Morris Report, was published in 1975 and was more directly concerned with housing issues.[81] Entitled 'Housing and Social Work : A Joint Approach', it was intended to anticipate some of the problems likely to arise from the location of social work departments at the regional level and housing departments at the district level after reorganisation. Accordingly, the Report stressed that unitary, comprehensive housing departments were likely to be a necessary prerequisite to ease of communication between the regional social work departments and the district housing departments over housing issues, such as homelessness. An important aspect of the Report's recommendations was that the housing directors should be given chief officer status to enable them to meet and talk with the regional directors of social work on an equal basis.[82] Finally, the other recommendations directly relevant to the housing service were that the training and selection of housing officers should be improved; obviously a reflection of the social work establishment's views on the calibre and attitudes of those in housing administration at that time. Summing up, the Morris Report followed the earlier Seebohm Report in pressing for improvements in the quality of the housing service through changes in the organisation and training of officers.

THE COMPREHENSIVE HOUSING SERVICE AND GROWTH OF HOUSING ADVICE CENTRES

Both the Seebohm and the Cullingworth Reports had pointed to a further important aspect of the comprehensive housing service idea. Not only did they advocate that local authorities should formulate their housing policies within the wider context of the local housing system but also that authorities should establish advisory services to assist those in housing need or difficulty. These housing advice centres were intended to deal with a wide range of clients with varying problems and opportunities in the local housing market. The idea of such advice centres was not entirely new. The Catholic Housing Aid Service had set up several centres during the nineteen-sixties, though these were based on a case work model,[83] while Seebohm and Cullingworth saw such centres as resembling Citizens Advice Bureaux.[84] Nevertheless, both reports envisaged that these centres would be primarily assisting the less well-off.

Despite the recommendations of the two official reports, the local authorities were slow in establishing centres. Instead the initiative was taken by organisations in the voluntary sector who built on the early work undertaken by the Catholic Housing Aid Centre (CHAC) The prototype of the new housing aid centres was sponsored by CHAC and Shelter and open in London during March 1970. The new centre, the Shelter Housing Aid Centre (SHAC), was intended as 'a place to which a person with any housing problem can come for help'.[85] It was not until several months later that the first local authority set up a housing aid centre. Perhaps not surprisingly, that first authority was Lambeth, the Director of Housing being Harry Simpson, a member of the Committee that produced the Cullingworth Report. After this lead many other authorities followed soon after with their own advice centres.

Both SHAC and the Lambeth Housing Advice Centre became important influences on the formation of these new HACs. The ministers in the Department of the Environment became enthusiastic converts to the idea of such centres - Peter Walker, the Secretary of State for the

Environment, visited both SHAC and the Lambeth HAC, and in 1971 Julian Amery, the Minister for Housing and Construction, sent a letter to those authorities regarded as being in areas of housing stress, recommending that they set up their own advice service and summing up the advantages of such centres for local authorities. This letter implicitly endorsed the idea of a comprehensive housing policy, though it did suggest that landlords should be able to make use of any advisory services.[86] The reason for the Conservative Government's surprising enthusiasm for HACs stemmed, at least in part, from their view that housing problems could be resolved if people were helped to participate more effectively in the housing market. They saw the barriers to this participation as being predominantly in people's lack of knowledge rather than their levels of income.[87] Similarly, many Conservative councils saw HACs in a similar light as 'widening the choice of housing available and relieving the pressure to build council houses'.[88] At the same time Labour authorities could see in the HAC idea a possible justification for further intervention in the local housing market and an instrument whereby their client groups, in the poorer sections of society, could be helped.[89] So the HACs could be perceived in a positive light by both the two main political parties despite their ideological differences.

Under these favourable conditions, HACs were established by many local authorities. By 1975 there were 54 HACs in London alone, though a number of these were run by voluntary organisations usually financed in part by local authorities. Outside London the growth of the centres was slower and these looked towards the London authorities in devising their own approaches. Of course, the Lambeth HAC and SHAC formed the models for the new HACs. SHAC, in particular, became a major source of advice, training and staff for the new HACs and one of the results of the sudden expansion was that in SHAC, between 1972 and 1975 there was almost a hundred percent turnover of staff.[90] At the same time a new occupational association grew up among staff working in the centres, though it has remained almost exclusively a London based association, the National Association of Housing Aid Workers. This Association has remained quite separate from the Institute of Housing. The latter has made little attempt to interest those working within the centres, while the housing aid workers themselves have not identified with the Institute. If anything, many of the housing aid workers were hostile towards the housing profession. They had entered the housing field for different reasons to those of most housing officers (at least of the older generation) and identify more readily with those in housing need than those involved in the bureaucratic tasks of rationing out the limited public housing available. Of course there is built into the relationship between a HAC and the housing department some potential for conflict over what constitutes priority cases.[91]

Housing Aid Centres have been generally recognised as one of the more important innovations in the local authority housing service. But the housing profession, although it was important in spreading the idea, did not originate the idea of such centres which rather came from the voluntary sector. As with the related idea of a comprehensive housing service, the housing professionals themselves were introduced to the new idea by official reports and the academic social welfare lobby. Once the local authorities began to establish the centres they still continued to rely heavily upon voluntary agencies to train and provide the staff. There was a related failure too by the professional community to attract and interest the new staff working in these centres. Despite some discussion at the time in the Institute about accommodating this new influx, no serious attempt was made to recruit these staff. One prominent housing professional observed at the time that 'the failure to absorb the housing aid service personnel into IHM ... and the emergence of a steering group to inaugurate a national housing aid association is a sad commentary on the Institute's inflexibility and lack of vision'.[92] Yet, on the other hand, conflict between the established housing staff and the new housing advice centre staff seems to have been almost inevitable as long as severe housing shortages persisted. In the event the housing aid workers formed their own association and went their own way.[93]

THE FIRST PROFESSIONAL PRESENCE IN CENTRAL GOVERNMENT

The first housing management adviser had been appointed in central government, in what was then the Ministry of Health, as early as 1944. No doubt this appointment was made in anticipation of the large house building programme intended by both parties in the Post-War period. Notably a member of the Women's Society of Housing Managers was appointed. Unfortunately, it has not been possible to discover how the adviser, Molly Empson, saw her role as few respondents could remember her, and even those who could, had only the haziest recollections. She appears to have been very much in the Octavia Hill tradition which was seen as rather dated by the sixties among those housing officers whom she met, but she seems to have enjoyed affectionate respect in housing circles. Just before her retirement in 1968, her post was advertised and Derek Fox, later to prove a controversial figure in the profession, was appointed.

Derek Fox began with the vaguest of job specifications. Although the intention had been for him to work with Molly Empson for several months before she finally left, she became seriously ill and Derek Fox only managed a brief telephone conversation with her. Derek Fox himself had serious doubts about whether he had the necessary experience for the post.[94] Until then his experience had been entirely within local government and he had taken no part in any national level consultations over housing policy. Nonetheless, he had spent over twenty years in local authority housing, and immediately before his appointment he had been the Deputy Director of Housing in Hounslow London Borough for three years.

One of the first tasks of the new adviser was to act as a joint secretary of the CHAC Housing Management Sub-Committee which was working on the Cullingworth Report. Although the Report had rejected any idea of increasing central control over local authorities as a means of persuading them to raise their standards of housing management, it did recommend strengthening the housing management advisory functions of the Ministry of Housing, primarily as a means to the dissemination of good practice. The Report contrasted the situation in housing management, where there was a single adviser, with that in related fields, in particular the Ministry had a hundred and fifty professional civil servants dealing with actual house building.[95] However, this disparity was allowed to continue for several years until the next Labour Government went some way towards correcting it. It should be noted, however, that this imbalance reflects, and indeed was the product, of the still dominant convention that the management unlike the construction of housing is entirely a matter for local authorities.

It would be fair to say that Derek Fox, in common with the more progressive housing officers, regarded the Cullingworth Report as a professional manifesto for the future of the housing profession. Accordingly, it became the basic remit for his job, which he saw in terms mainly of spreading and encouraging good housing management practices in local authorities rather than as advising the Ministry. Consequently, he spent much of his time travelling around the country, a daunting task in the days before reorganisation when there were more than 1400 housing authorities. At least partly to reduce this problem, he encouraged the formation of discussion groups among chief housing officers in particular counties. Through these groups, too, he was able to sound out officers on the major changes facing local government at this time, such as local government reorganisation (the possible threat posed by any reorganisation proposals must have further encouraged housing officers to join together), the introduction of 'fair rents' and the proposal that housing authorities should assume responsibility for homelessness from the social service departments.[96] But Derek Fox found himself facing a conservative group of officers:

> I saw it as my task to stomp around the country like a Billy Graham missionary, preaching with religious fervour (that, at least, was how my efforts were described by Institute of Housing Management Council members) to try to convert the non-believers. But before long I became a heretic, spending an increasing amount of time trying to persuade orthodox adherents of the profession that they needed to adapt to changes in society and to adopt new practices.[97]

The prime way in which Derek Fox attempted to stimulate discussion through these groups was the distribution of a series of 'Housing Management Advisory Papers' which covered issues like the selection of tenants, maintenance organisation, rent collection, the housing of the elderly and tenancy conditions.[98] These papers were intended as an 'attempt to guide and stimulate local authority housing managers and to provide a short-cut so that they can isolate national issues and identify solutions to local problems - not only now, but more importantly for the inauguration of new councils which will come into existence in the 1972 Act and take power in 1974'.[99]

It is clear from these papers that Derek Fox took his lead from the Cullingworth Report and that he saw himself elaborating on the recommendations of that Report. It is difficult to assess the impact of the Report on the housing managers in local authorities at this distance in time. Of course, Derek Fox was severely limited by the lack of any research or advisory support, he had to write and publicise the papers himself. They were widely regarded at that time as too progressive in local government housing circles where Fox himself, as he says, was widely regarded as a 'heretic'. But he appears to have achieved even less acceptability among the civil servants in the Department. Although many of them had a grudging respect for him as a progressive housing officer and, of course, Cullingworth had become highly thought of in the Department, Derek Fox was referred to by a couple of civil servant respondents as a 'maverick'. So, even though many of his civil service colleagues were in agreement with his views, including his critical view of current housing management practices, his direct, abrasive style and outspoken approach brought him into conflict with the Department.[100] Although his papers were subject to much less departmental scrutiny than those of the later housing advisors, there were still tensions between him and the Department over the extent to which he could act as a free agent. In retrospect he doubted whether the 'Department was ready for my vigour'.[101] He was also critical of the dominant approach of the Department:

> You see, all the best brains at DoE are almost full-time employed on baffling others who have only a little less knowledge on what is actually happening in the examination of housing policy (it's a game called Parliamentary Questions, although there is a variation called consultation).
>
> Eventually comes a time when these people, who should be dynamically deciding policy (and guiding it, monitoring it, adjusting it, even all-changing it), become so entrapped in their forced order of priorities that they even believe in it themselves: or else they become so cynical that they won't believe in anything - not even housing.[102]

There were also difficulties between the housing advisor and some regional offices. On a couple of occasions he had been horrified by the type of housing development that he saw taking place and had expressed these views forcibly to the local authorities and the regional architects concerned. The architects had responded that they were simply following ministerial instructions in giving approval to as many new housing projects as possible, a response that failed to satisfy him. They were further angered that Derek Fox had expressed his views directly to the local authorities without consulting the Department, which placed the architects in the regional offices in an embarrassing position. These complaints were passed on to headquarters, where his superiors were unsympathetic. These incidents further worsened his relationship with the Department, especially as he felt strongly that high density building was creating further problems for the future.

THE CHANGING APPROACH TO HOUSING POLICY: THE LABOUR GOVERNMENT 1974-79

The advent of the new Labour Government was to have major significance in the development of the housing profession. Tony Crosland was appointed Secretary of State for the Environment and Reg Freeson Minister of State for Housing within the Department. Both these appointments were significant. Tony Crosland was a champion of the interests of the working class as consumers and less in sympathy with those in the Labour Party who advocated a paternalistic-bureaucratic type of socialism. He was very much aware of the need for Labour to appeal to its supporters in these new terms.[103] More specifically, he was also anxious to change and reorganise the basis of housing finance which had become so complex and, to a Labour Minister, so anomalous in its redistributive consequences. Crosland's junior minister, Reg Freeson, came from a rather different background to that of his boss. Reg Freeson had entered the House from a background in local government, having been leader of the former Willesden Metropolitan Borough, which had given him first hand experience of the problems of housing at grassroots level.[104]

Not only had the individual actors changed, but the premises of the housing debate had also begun to change. In the two general elections of 1974 party competition over house-building figures was much less prominent than it had been at earlier elections. The debate over housing had shifted towards other issues, in particular the two main parties stressed their commitment to owner-occupation and competed over the concessions that they were willing to offer to prospective house buyers. The Labour Party Manifesto contained a pledge to protect council tenants by giving them security of tenure and there had been extensive discussions within the Party on tenant participation.[105] Nevertheless the Department had already become committed to reviewing housing management practices under the outgoing Conservative Administration which had commissioned a research study into existing tenant participation schemes.[106]

Many civil servants in the Department shared, too, the view that the conditions of council tenancies needed reform. The former Permanent Secretary in the Ministry of Housing had commented:

> The council tenancy, which historically was provided to help the poor and still has that dominating motive, has inevitably the implication of dependency - of a privilege being conferred instead of a choice being made .... Management, especially of the bigger municipal estates, goes by rule - as it must do where elected representatives are directly responsible. There is more scope for treating applicants or tenants individually.[107]

It is worth noting how this former, distinguished civil servant attached the responsibility for the restrictive nature of much housing management to the elected members. In the same paper, she went on to argue that local authorities should be deprived of their house building and most of their housing management responsibilities.[108] Certainly there was a widespread view within the Department that too many local authorities were failing to live up to their responsibilities to their tenants and that the Department itself had neglected management problems. At this time, too, the Department was beginning to become aware of the problem of difficult-to-let housing - the legacy of past housing policies that had concentrated on the numbers rather than the quality of housing. Difficult-to-let housing was seen as representing a considerable financial loss to the public purse and, less importantly to many Departmental officials, a serious social problem:[109]

> ... by this time the feeling in the Department was that housing management was as important, if not more important, than building new houses and improving old ones. It was as important to look after the existing investment, which was originally largely funded by the Department, as it was to look into the building of new houses. The background to the previous national housing policy had been the shortage of housing, by the mid nineteen-seventies that had been largely overcome. This new interest (in management) was especially strong after the difficult-to-let problem arose.[110]

So, when the new Labour Government entered office, there was a growing interest in the Department in questions of housing management. But by this time the housing advisory functions within the Department were non-existent. Derek Fox had left to become Director of Housing in Hammersmith and the Central Housing Advisory Committee was effectively defunct. The last two reports published by CHAC had been the Cullingworth Report and the Cohen Report on Housing in New Towns. After the latter Report was published in 1971, the Committee met very infrequently. This may have reflected the Conservative Administration's change of approach to housing compared with th Labour Government's and, as important, the Department's preoccupation with that major piece of legislation, the 1972 Housing Finance Act. For the first time this Act took the power to fix their own rents from local authorities, which made it one of the most controversial pieces of housing legislation ever considered by Parliament. The Conservative Ministers appear to have been suspicious too of CHAC feeling, as one civil servant put it, that 'its views did not always chime in with those of the Ministers'.[111]

The new ministerial team saw little point in reviving the Committee. Reg Freeson had not been very impressed by its workings when he had chaired it during the previous Labour Government, although he had considerable respect for certain of its reports.[112] Besides it was not the most appropriate instrument for the main task that Crosland had in mind, a review of housing finance. In 1974 he appointed a Housing Policy Review Panel to produce recommendations on this issue.[113] This Panel was dominated by Labour politicians and sympathisers, though there were several more neutral members. There was only one housing officer on the Panel, Harry Simpson, who by this time, had been elevated to what was the top job in housing, the Director of Housing in the GLC. Another early act of the Labour Government was the repeal of the Conservative Housing Act which it had so vigorously opposed in opposition.[114] This was done in the 1974 Housing Act which also abolished CHAC.

As far as the post of housing adviser was concerned, this had been left vacant after Derek Fox's departure. Before the election of the new Labour Government, a small group of civil servants within the DoE housing directorate had already been reviewing alternatives for a new housing advisory service. The interest of the new ministers gave them the necessary support to continue with this review.[115] The Department commissioned a past president of the Institute of Housing, Charles Webb (formerly Housing Director of Southend), to review the desirability of appointing a housing management adviser. Not unsurprisingly, given his professional allegiances, Webb came out strongly in favour of such an appointment. However, this report remained on the table. The new minister was involved in piloting the 1974 Housing Bill through Parliament and the officials in the Department had 'genuine doubts about such an appointment'.[116] Perhaps they were wary about introducing another, possibly abrasive personality into the Department. So it was not until the next year that further moves were made in the direction of establishing a new housing management advisory service within the Department.

Before the establishment of this service is discussed in the next section, mention should be made of a further development in housing policy. This development, which took place later in the life of the

Labour Government, was the introduction of the Housing Strategies and Investment Programmes (HIP) system. This was proposed in the Government's Green Paper on Housing policy and instituted in 1977.[117] The intention of the system was for the Department to determine authorities' housing allocations against the background of their local housing problems and proposed policies. As such authorities were expected to relate their local policies, especially those involving capital expenditure, to local conditions and to the contribution made by other organisations towards meting housing requirements.[118] This new approach to public housing finance was clearly an extension of previous initiatives intended to encourage local authorities to take a broad view of and responsibility for local housing conditions. As such it gave further support to the idea of a comprehensive housing service within which the HIPs submission would be drafted.

THE ESTABLISHMENT OF THE HOUSING SERVICES ADVISORY UNIT

It was finally decided to appoint a new housing adviser in 1975.[119] The new post was advertised at the equivalent level of senior principal, despite the view expressed by housing officers at that meeting that the Department would not get someone with the type of experience they wanted unless they advertised the post at a salary comparable with that of chief officers in the larger urban authorities. For this reason, then, the Department failed to attract the type of candidates which they thought suitable. Subsequently, the Civil Service Department was persuaded to raise the status of the post to the equivalent of assistant secretary but, as an interim measure to avoid delay while the negotiations and subsequent re-advertisement of the job were taking place, an existing housing director was persuaded to join the Department as the adviser on part-time secondment from his employing authority. The housing director was Bill Woof, Director of Housing in Gateshead, and was already well-known in the Department through his work as an AMA adviser. He seems to have been approached for several reasons. The Department already knew him both on the official and the ministerial sides and the arrangements for a secondment were probably made easier by the fact that his Housing Chairman was also the Labour Deputy Leader of the AMA at the time.

Bill Woof, joined the Department in May 1976, on the basis that sixty percent of his time was to be spent on DoE business. It was significant that his title was Housing Services Adviser, not Management Adviser. This reflected Departmental recognition of the principle of comprehensive housing service. Apart from his involvement with the new Housing Services Advisory Group, his other main task initially was to draw up a brief for a prospective Housing Services Advisory Unit. Preparing and agreeing a brief for this Unit turned out to be a long and difficult task.[120] He began with quite an ambitious scheme for a core of advisers in DoE headquarters with an adviser in each of the nine DoE regional offices. This scheme encountered considerable opposition within the Department. Firstly, there were financial objections. Secondly, there was some opposition from the regional controllers in charge of housing - placing officers within regions who had a remit that so closely involved local authorities, yet who must inevitably be junior to the regional controllers, was seen as raising many problems. The memory of Derek Fox's interventions too may have increased the wariness of the regional controllers. Thirdly, there were union objections. The Institute of Professional Civil Servants, reflecting the interests of the built environment professions - architects, surveyors and engineers - were unhappy about the introduction of a new professional group within the civil service which would increase the competition for senior positions and threaten their dominance in the housing areas of DoE's activities. Thus the Department had to prove to the unions that no one else in the Department could do the new advisers' jobs.

The final brief for a new Unit was not announced until the end of 1977. The views of the local authority associations and other interested parties were requested on which aspects of the housing service the unit should first direct its attentions.[121] The local authorities, on the whole, welcomed the idea of the Unit, the ADC was more enthusiastic about the Unit than it had been about the Group. In 1978, the post of Director of the Unit was advertised twice in the first half of that year. On the first attempt, after a trawl within the Department and advertising it for general competition, the Department had been unable to find the type of person they thought they required, so the post was re-advertised. Bill Woof applied on both occasions as he had developed an interest in and a commitment to the job. It seems very likely that the Department's hesitation in making a choice stemmed in large part from their having to appoint to a senior civil service post from among a group of outsiders, who had not developed the traditional skills and style of the civil service. Thus Woof himself was regarded as less able at drafting the conventional civil service report, but this was compensated for by what they saw as his main quality: 'his ability to get on with outsiders and Ministers, he instills confidence in Ministers of both parties'.[122] The same civil servant went on to describe Woof's eventual appointment, with hindsight, as a successful and 'inspired one' on his part and that of his colleagues.

Once Bill Woof was finally appointed, on a five year contract, the two other advisory posts which had been agreed within the Department were advertised. Unfortunately it is not known how many or what type of officer applied for these posts, but the Department found it easier this time to find two people who were acceptable. Significantly, the two officers selected were both prominent within the professional community, but they represented different branches and views within that community. Mary Smith, who joined in August 1978, was a chartered surveyor who had previously been the Chief Housing Manager of the Crown Estate Commissioners and her past career had been both in housing associations and in the London Boroughs. She was regarded within the housing profession as being within the tradition of the Women's Society. She was also a member of the IoH Council and the main author of the 'Guide to Housing', published by the Housing Centre Trust. In contrast, the other new adviser, Chris Gittins, had worked solely within local authorities, and had been the Director of Housing and Environmental Health in North Bedfordshire District Council as well as an ADC adviser. Unlike Mary Smith, he came from the more business efficiency orientated background of the old Institute of Housing. Gittins also had the advantage of being an environmental health officer so he could be called in for advice on improvement and rehabilitation policies in the absence of any existing environmental health advisory staff in DoE. Between them then, the three officers within the Unit combined a fairly comprehensive range of experience in the housing field: Bill Woof from a large metropolitan district, Mary Smith from housing associations and London Boroughs, and Chris Gittins from small, rural districts. This was intended by the Department for political reasons - so that every type of authority felt that it was represented in some way - and so that the Department had a large spectrum of advice on which to draw.

The original intention had been for the Unit to work closely with the Group, the latter acting as a steering committee. However, it did not work out in this way. The Group and the Unit went their own ways and over time became involved in different activities.[123] This is not to say that they have been completely separate. Even though only Bill Woof was formally a member of HSAG, the other two advisers normally attended Group meetings and worked on the HSAG sub-groups which were producing reports. And, of course, the main activity of the Group was the production of these reports. It should be added that the Group was serviced by one of the DoE policy divisions, H11, and not by the Unit.

The role of the Unit in DoE was greatly affected by the major piece of housing legislation being prepared within the Department, the Labour Housing Bill,[124] and by other Government housing initiatives. In fact, one of Bill Woof's first tasks was to give practical advice on matters such as drafting the Code of Guidance that accompanied the Homeless Persons Act 1977.[125] So the advisers were an important source of advice on the likely local implications of proposed legislation and central guidance such as circulars. Having advisers within the Department had the advantage that they could be consulted informally, often during the earliest stages of policy formulation, without having to go through the formal channels of the local authority associations in order to obtain advice on the problems of implementing policies at the grassroots level. This also meant that it was easier for the Department to maintain confidentiality for its deliberations.

Nevertheless, the advisers were not solely concerned with advising at the centre. They also spent a considerable proportion of their time, between thirty and forty percent, travelling around the country and visiting authorities and DoE regional offices.[126] The three advisers divided up the country into three sections - Bill Woof dealt with the North West, the Northern Region and Yorkshire and Humberside; Mary Smith with the South East, the South West and Greater London, and Chris Gittins with the Eastern Region and the East and West Midlands. Within their areas of the country, the advisers were active in attending meetings, usually IoH branch meetings and chief officer associations, and in paying visits to individual local authorities - their object was to visit all local housing authorities by the end of 1981. The visits to individual authorities arose in a variety of ways. Sometimes authorities requested a visit because they had some problem on which they were seeking advice, others issued an invitation because they wished to draw attention to some innovation or change that they had introduced. Yet again, in the case of some authorities, the advisers might have heard that an authority had some problem and would attempt to engineer an invitation. But Bill Woof was careful to stress that 'our involvement is by invitation ' not by thrusting ourselves upon local authorities'.[127] Their visits to authorities were intended as purely informal occasions. They were usually arranged informally during meetings at the Institute, at its branches or, most likely, at meetings of the chief housing officers' associations. These informal occasions gave the advisers plenty of opportunity to raise the possibility of a visit without the appearance of it being a central government tour of inspection. Clearly, the trust of other housing officers could only have been expected to be sustained for as long as the advisers continued to be seen as essentially local government housing officers working in central government.

The form of these visits varied considerably, the adviser meeting with whomever the authority decided to field. This could mean just meeting with the chief housing officer or, on occasion, even being presented to the whole housing committee; these arrangements could vary from visit to visit, depending on different authorities' styles and the types of issue that the adviser had expressed a particular interest in discussing. The Advisory Unit used to agree a list of topics that were to be covered with each authority visited, so this list provided the basis for discussion. The use of a systematised list in this way enabled some comparisons among authorities to be made and the visiting adviser to probe for problems. The topics on the list included relations with the social services departments (especially over homelessness), difficult-to-let housing and problems in maintenance organisations. Back in Marsham Street, the advisers wrote up reports on the authorities visited. These reports were distributed to the interested policy divisions in DoE and went to form the collection of information that the Unit was building up on housing practices. One adviser referred to this body of information as constituting part of a 'central information service' for housing authorities. This was another aspect of their work, as they saw it - the facilitation of information and idea exchange among authorities.[128] At the same time this collection of views and information from local authorities must have been of benefit to the rest of the Department concerned with housing. But it is difficult to give any estimate of the influence that this feedback had on the Department. The advisers themselves said that this feedback did have an effect on policy but neither they nor the policy divisions were willing to talk about specific examples.

The two new advisers had only been in post for just under a year when the new Conservative Government came into office in May 1979. The Government had a strong mandate to reduce public expenditure and the new Secretary of State for the Environment, Michael Heseltine, immediately set about reviewing the staffing of the Department. After this review, called the Ministerial Information System (MINIS), Heseltine announced that he intended drastically reducing the number of divisions within DoE.[129] One of the areas of departmental activity at which he was looking for cuts was housing management, which, it was argued, should not be a concern of a central department as it was solely a local responsibility. As a result of this review the housing policy command was reorganised and the policy division responsible for housing management matters, H11, was abolished in April 1981. However, although it was widely assumed that the Unit was also under threat, in the event it was reprieved and has continued in existence until the present (the end of 1981).

RECENT DEVELOPMENTS IN HOUSING TRAINING

The Department had identified training as a major priority in drawing up its plans for improving local housing management. The first HSAG report was on training, and previously the Department had commissioned research on the state of housing training from the City University. The civil servants were particularly struck by the small proportion of housing staff who had any qualifications at all, and the very small number who had the housing diploma awarded by the IoH. At that time, too, there was nothing like the number of housing students in training necessary to transform the numbers of trained staff in local housing departments.[130] This concern was reflected in a major ministerial statement by Tony Crosland:

> One of the major problems facing housing authorities and associations is the scarcity of qualified professional staff in this field. It is my intention to consider what positive help the Department can give in order to remedy this and to increase the opportunities for training and education for housing management. A fresh initiative on this front must embrace not only those who will reach the highest position in the housing service, but all who work it, at whatever level.[131]

In the event the HSAG Report took the view that training was crucial in raising the standard of housing management; an emphasis stronger than some politicians would have given.[132] The Group maintained that the absence of a tradition of training:

> ...together with the low status long accorded to the profession has discouraged recruitment of people of the right potential or attitudes. The importance of housing work has not been recognised, and it has remained the 'cinderella' of local government services.[133]

Under the prevailing conditions they found that little was being spent on housing training, and staff at all levels were often too hard pressed to have time for training activities. Notably, the Report linked this emphasis on training to improving officers' appreciation of the social context of housing work and of the social purpose of housing.[134]

To overcome these deficiencies a range of training initiatives were recommended. Particular significance was attached to the introduction of in-service training for all levels of staff, to improve the basic and, where necessary, more specialist housing knowledge of the lower grades and to give management training to the more senior staff. For those entering the profession, they proposed that the Institute diploma should be upgraded to degree level and arrangements be made for their release for training. To promote these measures in individual local authorities, they recommended that each housing department should have a part-time training officer. But the main recommendation, and the most controversial, was that a Central Housing Training Council should be established 'to oversee and co-ordinate this new drive in housing training'.[135] It would have involved assessing training needs, setting training standards, assisting local authorities in developing training programmes, acting as a focal point for exchanging training information and initiating research. But it was not intended to take over the examining functions performed by IoH. Rather the Group accepted the IoH argument that 'the conduct of examinations is central to the role of a healthy professional body',[136] a principle for which the professionals fought hard in the Group.[137] And, as it emerged from the Institute that they were endeavouring to raise the standard of their diploma, the Group decided not to recommend a major, and possibly unnecessary, change in the examination arrangements. Finally, they asked the government to contribute one million pounds over the next two to four years towards the cost of the Council and related training initiatives, including the payment of student training grants. It is worth noting that some of the housing officers on HSAG had favoured a housing college for senior staff, the idea apparently being mooted in conjunction with the Housing Centre Trust.[138] This was symptomatic of the attitude of many of the older and more senior housing directors who were anxious to distance themselves from the street-level housing management officers and emphasised their strategic policy role in terms of the Cullingworth Report.[139]

The HSAG Training Report was influenced both by the existing professional model of social work and the views of the City University Training Group.[140] It had drawn on the preliminary findings of the City University Housing Education and Training Project, the findings of which had been made available to the HSAG in advance of their publication. The first of the results of the Project were, in fact, published several months after the HSAG Report in November 1977.[141] Of the report itself, several respondents, especially in the Department, remarked that it was more important in influencing policy on housing training than the earlier one had been. Certainly, it was more firmly grounded in fact, based as it was on two years research, which gave its recommendations some weight.

The Report is valuable in this discussion of housing training because it gave a good picture of the state of training as it was in 1976, in both local authorities and housing associations.[142] The most striking finding was the low proportion of housing staff that held the housing diploma or any of the exemptions to it. Thus, of 34,000 non-manual staff working in housing departments, less than 3.5% were qualified. Not surprisingly, perhaps, given the history of the profession, of staff working within housing departments a rather higher percentage held qualifications other than the housing diploma than held the diploma itself; 1% belonged to CIPFA or the Environmental Health Officers Association (EHOA) and a further 3% held other qualifications.[143] Moreover the general level of educational attainment among housing staff was low.[144] This seemed to confirm the civil servants' worst suspicions about the calibre of housing staff.

There were few signs of improvement, according to the City University team, with the number of qualified staff only increasing slowly. In 1974, 136 officers gained the diploma, in 1975 only 43 qualified and, in 1976, 171, though on top of this there were about 20 students graduating each year from the housing management degree courses at Sheffield and Bristol Polytechnics which had begun in 1977 and gave their graduates full exemption from the examination requirements.[145] This low level of qualification reflected, at least in part, the fact that 'many staff, particularly those from the (housing) associations hold the Diploma to be of doubtful relevance in its present form'.[146]

The City University felt that it was a matter of urgency for the IoH to improve the level of its qualification,[147] a view that must have influenced the earlier HSAG Report. Of concern as well, was the high turnover of housing staff, as many as 30% had been in post for only a year or less.[148] Again, this high turnover and the low level of educational attainment were linked to the poor career prospects of housing officers. Interestingly the City University rejected any suggestion that training should become a national responsibility:

> ...we are specifically not saying that provision of training is in general a task that should be carried out at the national level. The current training activities of national bodies should be expanded and improved to meet those types of needs best dealt with at this level but the main activities should be performed locally.[149]

It seemed to them that the variations among housing organisations and the differences among the types of job, rendered a 'detailed national prescription for training' impossible. Rather they felt that it was important that housing training be kept relevant to individuals' work both in their own eyes and in those of senior management. Any new training initiative was very likely to face resistance from its supposed beneficiaries unless it was closely tied to their practical requirements.

Nevertheless, they advocated the setting up of a 'Housing Services Training Development Council'. This Council was intended, among other things, to encourage senior staff to concern themselves with training matters, identify problems and disseminate information on training.[150] This was broadly similar to the HSAG recommendation for a Housing Training Council, though the City University stressed the need for training in situ rather more than the earlier Report. They considered other possibilities - the idea of giving the Housing Corporation some responsibilities for training or the Local Government Training Board taking over training responsibility. They rejected these alternatives. The LGTB was seen as having insufficient experience in the housing field. In this rejection of LGTB role in housing training they were in agreement with most senior housing officers who felt that such an arrangement would fail to give the profession the sort of recognition that a separate training board would do. It must also have concerned them that they would have little control over the activities of the LGTB but would, very likely, be effectively in control of any housing training council that might be established. The idea of giving the Housing Corporation training powers was also unacceptable to the Institute of Housing, since it would effectively have split the profession between the public and the voluntary sectors.

The two reports, from HSAG and from the City University team, encountered some hostility within the Department after they were published. One civil servant member of HSAG commented, 'the amount of opposition that the Report faced from within the Department was quite a shock to me'.[151] This opposition led the civil servants connected with HSAG to cease pushing the idea of a major housing training initiative for a while:

> (After the Report came out) the City University Project had been underway. So we came to lean more on what was going to come up from them, soft-pedalling on our own Report, although we knew what their conclusions were going to be.[152]

The opposition was from the financial side of the Department which was resistant to any proposals which appeared likely to increase public spending.[153] In this light the housing civil servants realised that it would have been tactically wiser to hold back the HSAG Report until they had a stronger case. Their hope was that the City's Report would give them a further opportunity to re-open the issue and also support their arguments with some hard data.

The proposals of HSAG and the City University team faced not only opposition within the Department but also within the local authority associations. The associations did not accept the rejection of the LGTB as the appropriate body to sponsor a new initiative in housing training. Neither did the associations welcome the idea of a central training council in housing, for reasons which will become clearer in this section. But the reports did stimulate a spate of activity concerned with housing training. One of the first organisations to take advantage of the fact that training was now, temporarily at least, on the political agenda, was the Association of London Borough Housing Officers (ALBHO).[154] ALBHO supported the idea of a housing training council and advocated the appointment of housing training officers in all the London Boroughs, as well as emphasising the importance of in-house training for staff whose job meant dealing with the public directly.[155] These and other recommendations of ALBHO, insofar as they related to London, were endorsed by the London Boroughs Association. A London Housing Training Committee was appointed and within a couple of years almost half the London Boroughs had appointed training officers.[156] This seems to confirm the reputation that the London Boroughs had for being progressive in the housing field; for example, they had also been first in setting up housing advice centres.

After their initial response to the HSAG and City University Reports, the ADC and the AMA set up a small 'Steering Committee' at elected member level to respond to these reports, together with the Local Authority Conditions of Service Advisory Board (LACSAB) and the LGTB. It included two members from each of the two associations and one representative from each of the two boards, again these were elected members; and the committee was chaired by a member of HSAG,[157] Frank Bushell, in his capacity as Chairman of the ADC Housing and Environmental Health Committee.

The officer's group attached to the members' steering committee conducted a brief survey of twenty-two authorities. Not surprisingly, this survey arrived at some of the same conclusions as the City University team - that there was a lack of commitment to both in-service and out-service training in housing departments, that there was little management training for senior officers and that there was a danger that the ranks of qualified staff would actually begin to shrink as older staff began to retire. But the Steering Committee, that is the elected members, insisted that 'the development of housing training should be seen as an evolutionary process'[158] and, significantly, warned that 'there should be no question of training for its own sake'.[159] Even so, they affirmed, that there was an 'urgent need for increased training provision but primarily through the individual local authorities themselves with each authority appointing a full-time or part-time housing training officer'.[160] They were objecting to the formation of 'yet another ad hoc body' which would have to be connected up with the other central organisations in the field.[161] To establish such an organisation would have taken time and delayed further initiatives. Meanwhile, the LGTB was already in existence and had resources and experience in local government training. It was also a cheaper option to the City University's proposals which were 'over-elaborate and unnecessarily expensive'.[162] The Committee concluded by recommending that a special committee of the LGTB should be established to oversee housing training,[163] and the LGTB allocated an extra quarter of a million pounds in 1979/80 to start this initiative. These recommendations were later agreed by the new Conservative Government and the two associations.

However, the professional housing community regarded these recommendations as a very poor substitute for the central council. Firstly, it appeared likely to deepen the split between the housing associations and the local authorities. The LGTB's remit did not of course extend to the housing associations, so it looked as if the result would encourage the National Federation of Housing Associations to extend their training activities further. There was already concern among the IoH Council members that they were beginning to lose ground in the housing associations as the Institute was seen as primarily a public sector body, and as the training role of the NFHA expanded.[164] Secondly, the local authority associations' proposals would have meant that housing training was dependent on other bodies for support in maintaining and winning resources for housing training. Any training body would be a pressure group for housing, in its absence other, more powerful interests might push housing training into the background. Thirdly, a central council would have enhanced the status of the profession. This was especially important as all the major professional groups in local government had a central council (for example CCETSW in social work and the Council of Engineering Institutions) or some other well sanctioned arrangement.

After the agreement of the AMA and ADC to the steering committee's proposals, the LGTB set up a Housing Service Training Committee to supervise the allocation of the quarter of a million pounds allocated. This new Committee included three representatives from each of the AMA, the ADC and the National Joint Council, two representatives from the IoH, officials from DoE, the Manpower Services Commission and the Department of Education and Science, an academic, a NALGO representative, a co-opted director of housing and a director of personnel and management services. The Committee was chaired by Lord Greenwood, then the Chairman of the LGTB (and, incidentally, President of the Housing Centre Trust) Thus, the Institute of Housing had, in effect, just three representatives on the Committee,[165] though a further housing director joined the Committee later. This was very unsatisfactory from the Institute's point of view. They were 'concerned that there are too many elected members on the Committee and not enough practitioners'.[166] This seems to imply that, probably correctly, they perceived the elected members as hostile to their professional claims. Neither were there any housing educationalists on the Committee, Tony Eddison, the academic on the Committee and the Director of the School for Advanced Urban Studies, not being primarily involved in housing training. In the Institute's view, too, the absence of any representatives from the housing associations seemed to increase the division in the profession; but later, as the result of pressure from the professionals, a representative from the NFHA was invited onto the Committee.

The Committee contributed funds towards senior management training, these monies being channelled through central organisations like the School for Advanced Urban Studies. But the largest proportion of these monies was allocated through the provincial councils to 'ensure the widest possible consultation on local authorities' needs'. In some cases local advisory committees have been set up between the provincial council and the local IoH branch, but in other cases the provincial council's training officers were left to organise those courses they felt were necessary. It should be added too that the funds were intended to be spent on all forms of housing and not just on the training of prospective and actual members of the Institute. However, the funds allocated for 1979/80 were renewed for the next two years which reflected a continuing commitment by the local authority associations to the importance of housing training.

The Institute did not give up the idea of a central training council.[167] Towards the end of 1979, the Institute sent a delegation to see John Stanley, the Minister of Housing, about training policy for housing, in particular to express concern at the possibility of funds for housing training being cut off. Stanley took the view that training was a matter for local authorities as employers and instead was more interested in getting the co-operation of the Institute in the consultation process leading up to the Government's new Housing Bill. In particular he wanted the Institute to arrange a series of 'consultative seminars' around the country during the passage of the bill in the next year. These seminars were intended to explain the provisions of the Bill, especially the right-to-buy provisions which were likely to prove extremely controversial. The IoH delegation went away with the impression that Stanley would give the idea of a training council serious consideration, after the current set of public spending cuts, as a quid pro quo for the Institute's assistance.[168] But since then the Government has not initiated any new developments in housing training.

CONCLUSION

Although the origins of housing management can be traced back to the nineteenth century, it did not begin to emerge as a recognisable occupation group within the public sector until the nineteen-thirties. In any case before this the stock of public housing was scarcely large enough to warrant the establishment of a profession. Once a large stock of public housing had come into existence the housing policy community began to take an interest in the professionalisation of the management of this stock as a major contribution to improving the service to the tenants. Pressure from the housing community on local authorities had little impact on housing management until the early seventies as local authorities were not easily persuaded of the need to professionalise housing management. The tasks involved in housing management were already divided up among the established professions in local authorities which were reluctant to surrender them to a new professional group. At the same time the elected members did not accept, in general, that housing management tasks were other than routine and, in particular, they were unwilling to relinquish much influence over housing and management issues.

It was not until the seventies that housing became an important local government profession. There are two aspects to the quite rapid changes that took place during the decade - one was the spread of managerialism in local government and the other the new central government interest in local housing services. 'Managerialism' is used here as a bucket concept to cover the various aspects of the new interest in management and organisation that characterised local government in the sixties and seventies and influenced local government reorganisation. It covers the corporate planning and management movement and the other related fashions of that period. Of course it is not accidental that interest in the problems of management and co-ordination emerged at this time. The size of authorities was increasing rapidly as public expenditure rose and reorganisation created still larger authorities. These changes led to a search for both new techniques to cope with the management problems and for new professional accounts as professionals in management roles sought to combine their professional claims with new managerial claims. In other words the local government professions became more technobureaucratic.

The housing profession was a major beneficiary of managerialism. The earlier professional account emphasising the landlord-tenant relationship and unified housing management was superseded, or rather incorporated into an account that drew on managerialism and on a new policy definition of the housing problem. Thus the idea of a 'comprehensive housing service' became the dominant account. This account could be justified in managerialist terms as a move towards a programme or problem based department or service, as well as conducive to improved co-ordination within the housing department and with other departments. Another, and related, aspect of managerialism in local government was its stress on the positive role of authorities within the local environment - in searching out problems and intervening to remedy them. The redefinition of the national housing problem as a series of local housing problems and the corresponding notion that authorities should become responsible for all the housing in their areas, fitted in well with this aspect of managerialism. In this way, managerialism far from being anti-professional, as many of its exponents claimed it was, in effect supported professional expansionism.

The second major factor in the further professionalisation of housing was the new central government interest in local housing services. Central government interest had grown up among the national politicians and officials within the Department of the Environment. Electoral considerations, pressure from the political members of the policy community and the personal interest of two ministers placed local housing management on the Government's agenda. These factors reflected increasing doubts about the success of a housing policy that had concentrated almost exclusively on new building rather than on the quality of the houses built. The politicians initiated a search for new approaches to the planning, design and management of the housing stock. This was very much in line with the current of thinking among officials in the Department. The Departmental view was that local housing services and, particularly housing management, had been neglected for too long by the Department. There were both financial and wider social concerns behind this view - the discovery of the difficult-to-let problem and the loss of rent revenue and tenant discomfort it implied, rising levels of rent arrears, the belief that housing management was still 'punitive' and insensitive towards the tenants, and a desire to prevent more 'difficult-to-let' housing from being built. Under broad political support, then, the civil servants had promoted several new policy initiatives - the Housing Services Advisory Group and Unit, and support for the further professionalisation of housing.

But by the late seventies housing, in common with other aspiring professions, was faced with a less favourable climate for professional claims at both the national and local levels. In contrast, the more established professions, even those like social work and planning which were of comparatively recent origin, benefited from the earlier climate which was more favourable to professional claims. But, for housing, this changed climate meant that there was little support for improvements in housing education and for the idea of a central housing training council. From within DoE the onset of public expenditure cuts meant that there was little support for the idea of a central training council despite the importance which the leading housing professions gave to such a council in terms of raising the status of their profession vis-a-vis the other local government professions. Similarly within the local authority associations a new and more critical view of professional claims, in general, had grown up which meant that they were unsympathetic to yet another set of professional claims. Housing was not the only profession to receive a setback to its ambitions during this period, environmental health had their claims for improved qualifications and graduate only entry refused at about the same time.[169] This change of view within local government not only reflected tightening financial restraint but also a sense of increasing disillusionment with the idea of professionalism. There was a growing body of opinion within local government that the professions had failed to live up to the ambitious claims to special competences that had been made for them and that professionalism on the ground had contributed to spiralling salary bills in authorities. There was, too, a vague feeling among, especially Conservative, elected members that they had supported the establishment of new public services but not the creation of large new bureaucracies.[170]

In the two later chapters on housing in this study, two further aspects of the growing central government interest in local housing services will be considered. The first is the Housing Services Advisory Group which was established to promote, indirectly, improvements in local housing services. The second is the development of the idea of a charter of tenants' rights that was intended to raise the status of tenants within society and give them added freedom and protection in public housing. The emphasis will be on the role of the professional housing community in these two policy initiatives which are of considerable importance in their work and in the achievement of their professional claims.

NOTES

[1] Marion Brion and Anthea Tinker, Women in Housing : Access and Influence, Housing Centre Trust, London, p.60.

[2] This section has drawn on Brion and Tinker, Women in Housing.

[3] Brion and Tinker, Women in Housing, p.69.

[4] Brion and Tinker, Women in Housing, p.75.

[5] Stephen Merrett, State Housing in Britain, Routledge and Kegan Paul, London, 1979, p.26.

[6] Merrett, State Housing in Britain, p.27.

[7] Interestingly, this anticipates many of the later experiments in dealing with difficult-to-let housing.

[8] Dennis Crouch, 'Roots of the Institute', Local Government Chronical, 15 June 1979.

[9] Ibid.

[10] Ibid.

[11] Central Housing Advisory Commitee, The Management of Municipal Housing Estates, HMSO, 1939, para.17.

[12] Ibid, para.17.

[13] Ibid, para.1.

[14] Ibid, para.9.

[15] Ibid, para.11.

[16] Ibid, para.17.

[17] Ibid, para.16.

[18] Ibid, para.49.

[19] For an account of Aneurin Bevin's reasons for using local authorities as the instrument of post-war public house building see Michael Foot, Aneurin Bevan 1945-1960, Granada, London, 1975; pp.69-71.

[20] These were given in the Dudley Report on the Design of Dwellings, Report of Design of Dwellings Sub-committee of CHAC, HMSO, 1944. This Report laid down minimum conditions; eg three-bedroomed houses should have three living rooms and an area of 900 square feet.

[21] Interview with former civil servant.

[22] CHAC, Management of Municipal Housing Estates, Second Report of the CHAC Housing Management Sub-Committee, HMSO, 1945.

[23] Ibid, para.21.

[24] Ibid, para.26.

[25] Ibid, para.10.

[26] Ibid.

[27] CHAC, The Selection of Tenants and Transfers and Tax Charges, the Third Report of the CHAC Housing Management Sub-commitee, 1949.

[28] Though the London Government Act 1963 did abolish the residential qualification in London, many boroughs evaded the Act by introducing points schemes that awarded points for years of residence.

[29] CHAC, Councils and Their Houses, Report of the CHAC Housing Management Sub-Committee, HMSO, 1959, para.138.

[30] Brion and Tinker, Women in Housing, p.79.

[31] Brion and Tinker, Women in Housing, p.180.

[32] Joan Ash, 'The Rise and Fall of High-Rise Housing in England', in Clare Ungerson and Valerie Karn (eds), The Consumer Experience of Housing, Gower, Farnborough, 1980, p.102.

[33] CHAC, Councils and Their Houses, para.4 (emphasis as in the original).

[34] Ibid, para.46.

[35] Ibid, para.12

[36] Ibid.

[37] Ibid, para.34.

[38] Which the Report listed as follows: 'advice at the planning stage on estate layout and house design, letting the houses, transfers and exchanges, rent collection and repairs, including responsibility for special services; eg central heating, lifts, general care of estates and their amenities, housing welfare and liaison with the statutory and voluntary social services'. Ibid, para.134.

[39] Ibid.

[40] Ibid, para.33.

[41] Ibid, para.134

[42] Ibid, para.40.

[43] Ibid, para.136

[44] Ibid, para.137.

[45] Merritt, State Housing, p.253.

[46] Ibid, p.259.

[47] Charles Baker, 'Housing Associations' in Mary Smith (ed), Guide to Housing, 2nd edition, Housing Centre Trust, London, 1977, pp.138-9.

[48] Ministry of Housing and Local Government, The Housing Programme 1965-70, HMSO, 1965 (Cmnd 2838).

[49] Ibid.

[50] E W Cooney, 'High Flats in Local Authority Housing in England and Wales' in Anthony Sutcliffe, Multi-Storey Living : The British Working Class Experience, Croom Helm, London, 1974, pp.151-80.

[51] Report of the Ministry of Housing and Local Government 1965 and 1966, May 1967, HMSO, (Cmnd 3282), para.73.

[52] Ibid.

[53] Ibid.

[54] Ibid.

[55] See history in Peter Hall et al, The Containment of Urban England, Allen and Unwin, London, 1973, Vol.1, Ch.3.

[56] John Thane, 'What the Pressure Groups should be Doing', Local Government Chronicle, 14 October 1977, pp.822-23.

[57] These comments are based on Housing Centre Trust annual reports and interviews with members.

[58] Patrick Rivers, The Politics of Pressure, Harrap, London, 1974, p.137.

[59] Des Wilson, 'Pressure Groups and Housing', Local Government Chronicle, 14 October 1977.

[60] Jane Morton, A Decade of Housing Advice, SHAC, London, 1981.

[61] For an example see Rivers, The Politics of Pressure, pp.141-6.

[62] In interviews and mentioned in the local government press.

[63] For example, Neil MacIntosh in Speech to Annual Meeting of Housing Centre Trust, 1979. (Author in attendance.)

[64] See John Edwards and Richard Batley, The Politics of Positive Discrimination, Tavistock, London, 1978, esp. Ch.2.

[65] Especially D Donnison, The Government of Housing, Penguin, Harmondsworth, 1967.

[66] It is worth quoting the White Paper that preceded the Act to underline the new role expected of authorities: 'The Government looks to the local housing authorities to be the main instruments of this policy in their area ... the local authorities must take the lead and drive the policy forward in each town and district. They will need better powers, better techniques and better financial arrangements'. Old Houses into New Homes, HMSO, April 1968 (Cmnd 3602), para.7.

[67] See, for example, John Greve, 'Housing Policies and Practices' in W A Robson and Bernard Crick, The Future of the Social Services, Penguin Books, Harmondsworth, 1970, pp.52-65. There were, of course, at this time, over 1500 local housing authorities.

[68] CHAC, Council Housing : Purposes, Procedures and Priorities (the Cullingworth Report), HMSO, 1969, Ch.1.

[69] Ibid.

[70] Ibid, para.468.

[71] Personal interview with Derek Fox.

[72] Report of the Committee on Local Authority and Allied Personal Social Services, HMSO, July 1968, (Cmnd 3702).

[73] The New Local Authorities : Management and Structure, Report of a Study Group on Local Authority Management Structure, HMSO, 1972.

[74] Ibid, para.9.33.

[75] Institute of Housing, 'A Comprehensive Housing Service - Organisation and Function', August 1972.

[76] In a personal interview.

[77] In a personal interview and, more officially, 'The Committee were concerned that many of the councils likely to be merged in the formation of the new districts will have had little or no experience of a comprehensive housing service and the departmental organisation necessary for such a service'. Institute of Housing Managers, Annual Report of Council, 1971/72, 1972. Notably the Working Party that drafted the Report was chaired by a director of housing from London, Harry Skidmore from Westminster.

[78] W Woof, 'The Comprehensive Housing Service - The Reality after Reorganisation', Housing Monthly, April/May 1975, p.7. This is the same Bill Woof who later became the Director of the Housing Services Advisory Unit.

[79] Woof refers to rent collection as a 'fundamental part of housing management (which) must surely be an essential function of any housing department'. Ibid, p.7.

[80] The New Scottish Local Authorities - Organisational and Management Structure, HMSO, 1973.

[81] Scottish Development Department, Housing and Social Work : A Joint Approach', HMSO, 1975.

[82] Ibid, para.2.8-10.

[83] Morton, A Decade of Housing Advice, p.5.

[84] Seebohm Report, para.391; Cullingworth Report, para.77-8.

[85] Morton, A Decade of Housing Advice, p.4.

[86] 'A housing aid centre can bring several important benefits. It will throw up much information about housing needs and conditions which might otherwise be missed. It should lead to action being taken to cater for the total needs of the areas and to solve the housing problems of various sorts of people, including landlords. It will lead to the best use being made of our stock of rented housing,both public and private. In fact, it will save an authority money by sometimes finding solutions that do not involve putting people on the local authority's housing list, for example by helping people to move to other areas or to buy a house of their own. In such a centre, questions of improvement grants and slum clearance proposals, landlord and tenant matters, house purchase, all of which can be perplexing to ordinary people, can be explained in simple and homely terms. All this will go to show that local authorities are manned by humane and understanding persons who grasp the human problems of those they serve.' Quoted in Derek Fox, The Selection of Tenants, DoE, August 1972.

[87] Cf Morton's comment, 'To some extent, the Government's enthusiasm was based on a hope that was not in the event to be borne out - namely, that the centres would demonstrate that the housing market was self-regulating.' Morton, A Decade of Housing Advice, p.8.

[88] Michael Harloe, Richard Minns and Jennifer Stober, Housing Advice Centres, Shelter/CES, October 1976, p.21.

[89] Cf 'At the local level, the Seebohm Proposals, combined with the now widely accepted proposals for a far reaching Housing Advisory Service in every area, should lead to a major advance in dealing with multiple deprivation', Tony Crosland, Towards a Labour Housing Policy, Fabian Tract 410, July 1971, p.10.

[90] Harloe et al, Housing Advice Centres.

[91] The organisation of HACs differs widely among authorities with varying degrees of independence from the local authority, in some authorities the HAC is run independently of the department

and in others it forms just a section within the department. Harloe et al, Housing Advice Centres.

[92] Derek Fox, 'A Case of Stunted Growth', Municipal Journal/ Municipal Engineering, Housing Management Supplement, 31 October 1975.

[93] These comments are more appropriate for London than the rest of the country, the AHAW being London-centred.

[94] In a personal interview.

[95] Cullingworth Report.

[96] These issues have been taken from the 1972 minutes of these discussion groups.

[97] Derek Fox, 'A Case of Stunted Growth'.

[98] The titles of the papers were as follows: 'The Selection of Tenants', 'Let's Abolish Residential Qualifications', 'The Local Authority Housing Maintenance Service', Rent Collection', 'Housing Needs of the Elderly', 'The Organisation of the Local Authority Housing Service' and 'Tenancy Conditions'. These papers were all distributed as typescripts during the IoH Conference in August 1972.

[99] Derek Fox, 'The Purpose of the Housing Advisory Papers', DoE, 1972, para.6.

[100] Personal interviews with civil servants.

[101] Personal interview.

[102] Municipal Engineering, 7 November 1975.

[103] See his A Social Democratic Britain, Fabian Trent 404, 1970.

[104] Personal interview.

[105] For example, a commitment to establish local tenants advisory committees had been in the 1959 election manifesto.

[106] By Anne Richardson.

[107] Lady Evelyn Sharp, Housing in Britain - Successes, Failures and the Future, 1970 Bellman Memorial Lecture, Abbey National Building Society, p.4.

[108] Ibid.

[109] Comment based on interviews.

[110] Personal interview with civil servant.

[111] Personal interview with civil servant.

[112] Personal interview.

[113] Originally established in November 1974 to report in one year, Local Government Chronicle, 22 November 1974. In the event it took almost three years.

[114] The Government 'will restore to local authorities the right to fix rents that do not make a profit out of their tenants'. Labour Manifesto.

[115] Personal interviews.

[116] Personal interview with civil servant.

[117] Housing Policy - A Consultative Document, HMSO, June 1977 (Cmnd 6851).

[118] See discussion of HIPs in chapter eight.

[119] The interviewees were uncertain precisely when.

[120] This section draws on personal interviews.

[121] In HPR/MAN (77)/2/October 1977, Annex 5, para.3.

[122] Personal interview with civil servant.

[123] 'I suppose there may be people who think that HSAG is the housing committee and the HSAU who are the officers serving that committee but it isn't that sort of relationship. HSAG, having a whole range, is now firmly established as a group that picks up an issue and then produces fairly concise but valid reports, which appear to have been well received'. Interview with Bill Woof in Local Government Chronicle, 7 April 1978, pp.366-7.

[124] See chapter nine for a discussion of the housing management implications of this Bill.

[125] Mentioned by another respondent.

[126] Personal interviews.

[127] Quoted in Municipal Journal, 22 June 1979.

[128] In spreading good practice, according to Bill Woof, 'we either broker this ourselves or put authorities in touch with each other' and 'we can give enough advice and experience to all authorities to avoid the bed of nettles that some of their colleagues have gone through already'. Ibid, p.666.

[129] In a speech circulated to DoE staff, July 1979.

[130] Derek Fox pointed out that since 1965 there had been an average net increase of only thirty qualified members of IoH - in 1977 only 77 candidates were awarded the Diploma. Municipal Engineering.

[131] Speech given by Tony Crosland, August 1975, quoted in Housing Services Advisory Group, 'Training for Housing Work', DoE, 1977

[132] This emerged in interviews with both sides.

[133] HSAG, 'Training'.

[134] '... among all grades the social purpose of housing is neglectedand there is a lack of appreciation of the "social context" of most housing work'. Ibid, p.15.

[135] Ibid, para.45.

[136] Ibid, para.18. It should be aded that it also ensures a secure and regular income for the Institute.

[137] Personal interviews with HSAG members.

[138] Interviewees mentioned this.

[139] One former president of the IoH became quite angry when the author called him a 'housing manager'.

[140] Later reported in City University, 'Housing Training : Findings and Recommendations', Education and Training for Social Work Project, Final Report, November 1977.

[141] City University, 'Housing Training'.

[142] The research was based on six housing associations and six local authorities.

[143] Ibid, p.8. By 1984, Derek Fox estimates that there will be 55 graduates from the three polytechnics running degree courses in housing (Bristol, Sheffield and Trent). Personal communication to the author.

[144] Only 15% held HNC/Diploma/degree or higher qualification, while 31% had no qualification at all and, of the rest, 10% had some post-school non-academic qualification and 39% had some kind of secondary school qualification. Ibid, p.19.

[145] Ibid, p.8. Note that this was the grand total of staff entering local authorities <u>and</u> housing associations.

[146] Ibid, p.7.

[147] Ibid, p.9.

[148] Ibid.

[149] Ibid, p.29.

[150] Ibid, p.47. The price tag attached to their proposals was £5 million.

[151] In a personal interview.

[152] Personal interview with civil servant.

[153] This was almost a year after Tony Crosland's 'the party's ended' speech in May 1975.

[154] 'Housing Training for London Boroughs', Report of Working Group on Training in Housing Work, ALBHO, February 1978.

[155] Ibid.

[156] Derek Fox, 'Education and Training for the Housing Service', Housing, March 19890, p.5.

[157] Though he had not been a members of the HSAG sub-group that had drafted the report.

[158] ADC/AMA 'Report of a Working Group on Housing Training', p.7.

[159] Ibid, p.7.

[160] Ibid.

[161] Ibid.

[162] Ibid, p.10.

[163] Ibid.

[164] This emerged during interviews.

[165] Diane Balfe, 'The Split in Housing Training', Local Government Chronicle, 19 October 1979.

[166] Ibid.

[167] 'We are still in favour of a central housing training council and we will continue to press for it. There are also doubts about whether the LGTB and the local authority associations will be able to raise sufficient funds. We will still be pressing for exchequer funds and a broader representation with appointments by the Secretary of State.' Dennis Crouch, Secretary of IoH, Local Government Chronicle, May 1979.

[168] Based on personal interviews.

[169] See 'Report on Staffing in Environmental Health', ADC/AMA/LACSAB/LGTB, April 1978.

[170] This position may appear rather naive in assuming that service delivery need not be associated with the creation of large sclae organisations to co-ordinate the service providers, but many local and national Conservative politicians do combine support for particular services with strident criticism of those not directly involved in service provision, especially health service and local authority administrators.

6 Highway agencies

INTRODUCTION

Problems over highway agency agreements between counties and districts arose in the wake of local government reorganisation. From chapter four it will be recalled that reorganisation divided municipal engineers into three camps, those in the counties, in the district councils and in the water authorities. The continuing viability of engineering at the district level then became partly dependent on agency agreements - with the counties over highways and with the water authorities over sewerage. If these agencies were to go from the districts, then engineers faced difficulties in justifying their continuing presence in many, and perhaps most, district councils. Although not all district councils had these agencies, almost all larger districts had them after 1974. The agencies, as will be seen, were particularly contentious in those areas which had previously enjoyed county borough status, partly because people in those authorities felt that their authorities were still large enough to exercise such powers and partly because, over the years, they had become used to exercising them. In the metropolitan counties the conflicts over agency were probably keenest. For the main executive function of the metropolitan counties was transportation and roads, so that when the metropolitan districts challenged the counties those in the latter felt their raison d'etre was at stake. In this chapter, the agency issue will be placed in its wider context, focussing on how it was faced as an issue at the national level and on the attempt, on a central initiative, to resolve it. But firstly, the historical background will be sketched out.

THE HISTORICAL BACKGROUND

Before local government reorganisation, a considerable number of district authorities exercised highway powers on a delegated basis from county councils. In the major urban centres the county boroughs were themselves the highway authorities for their areas. But in country areas the delegation of powers from the counties to the district and borough councils was fairly common in other services as well, such as planning, education, health and welfare. There were limits, in terms of population, on the size of district council which could apply for such delegated powers. Thus, to hold highway responsibilities, district councils were required to have a minimum population of 20,000 before they could apply as of right for the powers to maintain and improve unclassified roads; though they had no such right to demand delegated powers over other county roads. The Redcliffe-Maud Report, reflecting the new managerial thinking about local government, found the arguments against these delegated powers to be 'overwhelming'.[1] The districts were seen as too small for delegation to be economic and the relationship 'between two levels of elected authorities ... not compatible with their responsibilities to their respective electorates'.[2] Interestingly, the Ministry of Transport, in their evidence to the Commission, 'stressed the advantages, indeed in their view the necessity, of making a single authority responsible in each area for planning and transportation'.[3] Other official reports at the time similarly deplored agency and similar arrangements involving divided responsibilities, in particular the Marshall Report on Highway Maintenance.[4]

Despite the opposition in these official reports, however, the Local Government Act 1972 did permit district councils to apply, as of right, for powers to maintain unclassified urban roads.[5] This section had not been in the original Bill but had been included as an amendment during its passage through Parliament. The amendment was the result of pressure from the old county boroughs, who had extensive support in Parliament. The government accepted this amendment, which was hardly noticed amongst the more momentous changes taking place at the time, as a concession towards the county boroughs which were being shorn of their powers. Also in the Act the counties were made responsible for the formulation of structure plans, while the districts were to deal with the detailed application of those plans in development control. This division of powers was to create at least as many problems as the provision for agency agreements.

As far as other services were concerned, there was provision in the Act for agencies in the delivery of services, with the exception of education and social services but, unlike the case of highways, there was no provision for these agencies to be obligatory for the county councils. Although these other services were not seriously affected by agencies, the fact that they were allocated to one tier of local government rather than another remained controversial, long after reorganisation had taken place. As will be seen the controversy over these services helped keep alive at the national level the controversy over agencies.

The new non-metropolitan counties became responsible for education, social services, strategic planning (the structure plan, strategic development control and other powers), consumer protection, police, fire, some recreational activities and the disposal, but not collection, of refuse. The districts in these counties were left with one major service in housing plus planning (development control, local plans, etc), environmental health and other environmental services such as refuse collection and building regulations. While the pattern outside the major conurbations was for services to be concentrated at the county level, inside the conurbations services were concentrated at the district level. Thus, the metropolitan districts ran education, social services and housing as well as those services run by the non-metropolitan districts. This left the metropolitan counties with transport and highways, as well as their main executive function, together with strategic planning, police and fire, and some environmental services.

So the structural changes involved in local government reorganisation in England and Wales, created considerable potential for county-district conflict, both in the metropolitan and non-metropolitan areas. In the latter, the larger towns and cities resented the loss of services like education and social services, a resentment undoubtedly deepened by political party differences with the larger, and usually, Labour districts anxious to exercise power over services they saw as potentially redistributive. Meanwhile, the metropolitan counties were subject to challenge by the districts over the powers they exercised. So during the nineteen seventies pressure for the further reform of local government was kept up by the larger, non-metropolitan districts, especially by the 'Big Nine'.[6] They wanted a return to single-tier, county borough type powers for the larger districts. This pressure was exerted through the Association of District Councils and, more importantly, through political party contacts with Members of Parliament and, through them, ministers in the Labour Cabinet.

ORGANIC CHANGE

In 1974 the incoming Labour Government with its proposals for devolution placed the issue of the distribution of services among governmental bodies back on the political agenda. In brief, devolution arose as an issue after the growth of nationalist parties in Scotland and Wales which had gained some successes in the election in areas formerly regarded as Labour strongholds. This prompted the Labour Government to seek ways of defusing this new opposition. The response was to put forward Labour's own proposals for devolution in Scotland and Wales. The Royal Commission on the Constitution had also reported the year before and had outlined various possible schemes.[7] Although the majority report of the Commission had not come out in favour of extending devolution to England, the minority reports had and it seemed likely that Scottish and Welsh devolution proposals would be seen as linked to devolution in England. This was probably made all the more certain by pressure from Labour MPs representing depressed areas, like the North East, who felt that the creation of Scottish or Welsh assemblies would give those areas an advantage over their own regions in the competition for government money. At any rate, in 1974 the Government published a White Paper outlining some alternatives for UK devolution.[8] But it was not until two years later that the more substantial proposals for England itself were published in a White Paper.[9]

Early in 1977, the Prime Minister announced that the appropriate ministers would be asked to examine, with interested parties, the possibility of limited changes in local government organisation. The Association of District Councils responded by raising the issue of the existing division of functions between the counties and districts. They criticised the existing division of functions as leading to the expensive duplication of services and to some major services becoming remote from the people they served.[10] They stressed too that district authorities should be given powers to adopt an 'overall approach to the needs and problems of their areas',[11] especially in the large urban centres. Despite this last argument, the ADC did not press for major change. They advocated a 'limited review' within the present structure to remedy what they considered to be the defects of the existing setup. In particular, they wanted to see some of the larger districts having more control over three major local services - personal social services, planning and highways, and traffic management.

In July 1978 Peter Shore, the then Secretary of State for the Environment, made a statement in the House on what he called 'Organic Change'.[12] This statement contained some of the same arguments used by the ADC. He saw prospective changes as being 'greatest in the larger non-metropolitan towns and cities which would then be better able to deal with their problems, including inner city problems in their own way'.[13] He envisaged further discussions taking place over four main functions, personal social services, education, highways and planning. While on highways he held out some hope for the large districts acquiring highway powers.[14]

The County Surveyors Society responded quickly to this announcement. It was discussed in their Committee One, whose Chairman, the then County Surveyor of Merseyside Metropolitan County Council, drafted a report outlining the Surveyors' views.[15] This report was written in anticipation of an Association of County Councils request for advice and the CSS was anxious to have a prepared line for the county surveyor advisers to the ACC - a line that would reflect the views of the surveyors as a whole. Thus the Report was also intended to initiate further discussion among the membership of the CSS. Later the Report was circulated by the surveyors to their chairmen and other leading members as a background briefing paper on organic change representing, as it did, a mine of arguments supporting the counties' case. It is also interesting as a summary of the surveyors' views on agencies, although some saw it as too strong a statement of their views.[16]

The report outlined the dissatisfaction with the current agency arrangements from all points of view. It criticised district councillors for being dissatisfied because their own pet schemes were not approved and for failing to realise that the counties had to take decisions on a strategic basis and that they operated under severe financial constraints. As for county councillors, they were said to be discontented with the agencies because they wanted to control their own financial and transport policies, and had serious doubts about the efficiency of the district agencies, being critical of the delays resulting from two tier working. The view of the surveyors, in the Report, was that the abolition of the agencies would remove these sources of dissatisfaction among the members.[17]

As far as the officers were concerned in both district and county, the document took the vital question to be that of job satisfaction. Chief officers at the county level were frustrated because they had '<u>professional responsibility and accountability</u> for the performance of tasks and the provision of services, and <u>yet they have little or no control over work carried out by the Agent Authorities</u>'.[18] In the view of the county surveyors, the costing and measurement systems of districts were often weak and many of their direct labour forces lacked expertise and the county surveyors argued that district staff would be allocated more cost effectively if they came under the control of the counties. Finally, the district chief officers faced split loyalties and they were employed by the district, yet had to have some professional allegiance to the county.[19]

The Report also mentioned other problems arising for professional officers. More junior officers in the districts were apparently finding it difficult first to get suitable training for chartered status, and then to acquire the range of practical experience necessary for further promotion. This was presented in the Report as decreasing the attractiveness of a career in local government for new recruits.

Against this background, in the view of the County Surveyors, organic change could only be more divisive and lead to further inefficiencies and cost. Moreover, and this point was stressed, it would make the counties' comprehensive responsibilities in transportation and planning more difficult to exercise. The Report referred to the intentions behind the reorganisation of local government. The Secretary of State had, at that time, stated in a circular that the 'comprehensive responsibility' of the county council as a highway and traffic authority was linked with its function as the structure planning authority and with its duty to promote a co-ordinated and efficient public transport service. Furthermore, it was 'essential that this comprehensive responsibility remains intact'.[20] Though it would appear from Shore's July statement in the House that he was not intending to erode this comprehensive responsibility of the counties.

It is quite clear from this document that the county surveyors did not intend to consent to any major concessions in the direction of the districts. The type of changes apparently implied by the 'Organic Change' proposals were unacceptable to them. Their preferred solution to the problems involved in agency arrangements, which they of course saw differently from the district interests, was for the districts' right to maintain their roads to be withdrawn and for the control of the counties over bus undertakings and parking to be strengthened.

However, as a group, the surveyors were not all in favour of going for this preferred solution regardless. There were some serious reservations. The political climate had to be considered as it by no means favoured the counties - the Labour Government was contemplating shifting powers away from them and the districts' lobby in Parliament was strong.[21] This climate called into question the political feasibility of pressing for their preferred solution. At a time when regional government had become a live issue, those with an interest in perpetuating the counties did not wish to give any hostages to fortune by relying too heavily in the county-district debate on arguments that the bigger the organisation the better - that is, the more efficient and effective - because it seemed quite likely that the surveyors and other county people would soon be in the position of having to argue against new regional proposals from the Government. Many officers and members at the time saw the danger of the counties being seen as the redundant tier between the districts and a new regional layer of government. There was a more specific fear, too, among the surveyors. The counties maintained trunk roads themselves on an agency basis for the DTp, and the surveyors felt strongly that these arrangements involved too close a supervision of their work and they had been pressing the Department to loosen their control over agencies.[22] So, again, the surveyors were wary of using arguments that might later be used against their own trunk road agencies by the Department and, as was mentioned in the earlier chapter, they felt that their engineering colleagues in DTp were unlikely to give ground over trunk road agencies. Thus there existed in the Department a strong lobby which would seize on any opportunities to criticise the counties' stance. Their awareness of this professional civil servant lobby had an important influence on the way in which the surveyors argued their case in the later negotiations.

There was a further consideration which related to professional solidarity. After the districts had lost water and sewage disposal agencies to the water authorities, they were left only with agencies in for sewerage itself and highways, so that to remove highway agencies would seriously weaken, if not eliminate, municipal engineering at the district level. Some surveyors, especially the older ones, felt strongly about the need to maintain the representation of municipal engineering at all levels of government as they believed in the strength of a unified profession[23] and in the importance of decision-makers having engineering advice available. There was a more practical reason too, as one county surveyor put it, 'despite the problems of agency, I'd rather deal with a district engineer than a district planner or whoever'.[24]

The government's proposals for organic change were finally published in January 1979.[25] The White Paper accepted many of the districts' arguments, in particular those of the larger districts. It began by stating that many of the shire counties were generally regarded as being too remote from local people and noting that 'such criticisms are strongest in relation to the larger non-metropolitan districts, particularly the former county boroughs; these cities - which have a long and distinguished history of local autonomy - have resented the fact that they were deprived by the 1972 Act of many of the functions they had previously exercised'.[26] It maintained that the response to the earlier consultation document on devolution had demonstrated that there was considerable support for change. But the government did not intend to initiate a 'major upheaval'. The main area in which they envisaged change was in the personal social services where the government felt there was most scope for devolving powers to the districts. But on highways the Government did not propose radical changes. The Government accepted the broad principle, which the counties had been pressing, of the absurdity of separating the county responsibility for strategic planning and resource allocation from the actual provision and maintenance of the highways. Thus the road system should be 'planned, provided and maintained on a countywide basis in order to ensure the adoption of consistent policies, standards and priorities.'[27] Consequently the ADC proposal for the districts to be given responsibility for one type of road and the counties for another was ruled out as creating a potential for even more conflict.[28] However, the districts were not entirely disappointed. In the Paper, the Government asserted that 'as many as possible of the functions which affect people's lives on a day to day basis should be taken by the authority most involved with those concerned'.[29] The functions that Government had in mind were traffic management, pedestrianisation, pedestrian crossings and the provision of road signs. But, as the Government accepted, the operational logic meant that districts exercising such powers would have to have some powers to carry out auxiliary works on the roads. This obviously placed an important lever in the districts' hands. Significantly, the White Paper excluded the metropolitan districts from applying for traffic management powers on the grounds that every continuous urban area should have uniform traffic management practices. But, as far as the other districts were concerned, these powers would only be available to those that had an 'urban nucleus' of 50,000 population and a minimum population of 100,000.[30] This change would have encompassed many of the former county boroughs, not just the 'Big Nine'. Summing up, the White Paper represented a broad acceptance of the districts' case, though it did not go as far as the districts would have liked. The Government clearly had some difficulty in striking a balance between the demands of the districts and the need to give the counties a workable role. But the Government had reaffirmed the fundamental basis of the counties' case against the devolution of highway powers to the districts - that is the inseparability of strategic functions from the implementation of highways policy.[31]

The ADC response to the White Paper was lukewarm, while the ACC were clearly perturbed by the threat of the counties losing powers. The ADC singled out highway responsibilities for special attention as 'they are part of the very fabric of the community and directly impact on people's daily lives'.[32] Thus the lynchpin of the districts' case was that 'direct responsibility would give local democratic control over this local community service and restore a more intelligible and accessible link between the public and the highways authority'.[33] The direct conferment of highway responsibilities on the districts would have the further advantage of eliminating the problems of agencies, that is duplication of effort, wasted resources and lack of district control. So the ADC 'deeply regretted' that, in the Government's proposals, only traffic management functions were to be transferred. They disputed the Government's view that traffic management could be separated from highway improvement and maintenance. As the proposals stood too, only about sixty of the Association's members could have claimed these traffic management powers, so the districts wanted the criteria to be loosened to include a wider range of councils. Their point was that many of the districts excluded had demonstrated their suitability through their present agencies and their past experience as county boroughs.

The ADC did not concur either with the White Paper's view that the role of the county as a strategic planning authority in both land-use and transportation was inseparable from the provision and maintenance of roads. They instanced the fact that municipal public transport undertakings functioned with few problems within the context of overall county transportation policies. So the Association set out their own proposals for organising the relationship between a county with a strategic transportation function and districts with at least some highway responsibilities.[34] Their argument was that if the Government's transport policies could be separated from those of the counties, then district policies could be separated from those of the county. The policies of the former would be more long term and 'strategic'. The districts would prepare their own parts of county Transportation Policies and Programmes (TPPs) with an obligation to consult with the county and ensure that the policies were in accordance with those of the county, as found in the first half of a re-jigged TPP.[35] It is worth noting how the ADC seized on the principle on which the county-central government relationship was based - in formal terms at least - to justify their own stance. What is rather surprising about this particular point in the districts' response was that they still thought that the notion of a 'strategic' function was both tenable in practice and at all attractive to the central departments. Certainly the experience in the planning field was not encouraging. Moreover, the districts' proposals seemed to guarantee that district-county conflicts would be exported up to the centre, whereas the existing arrangement, giving sole responsibility to the counties for the TPPs, meant that conflicts had to be resolved at the local level. Presumably, this must have been an important consideration behind Peter Shore's affirmation of the counties' strategic function, and their sole responsibility for the TPP, in 'Organic Change' and in his earlier statement.

There is little point in dealing in detail with the ACC response to 'Organic Change' since it followed similar lines to those in the earlier County Surveyors Society report. The counties stressed that the major objective should be to get the most economical and efficient service possible, and only the counties were large enough to provide such a service. Again this was directed against the districts' argument that to ensure local democratic control, highway responsibilities should be located at the lowest possible level of government. In the next section the way in which the two sides attempted to resolve these differences will be discussed.

THE ACC/ADC JOINT WORKING PARTY

In May 1979 under the threat of government intervention to legislate on local government organisation, the ACC and ADC formed their own joint working party at officer level, to reach agreement.[36] They had some terms of reference before them which had been drawn up by a joint elected members group. However these terms of reference seem to have been of little importance in defining the issues at stake more tightly than before. The county surveyors were immediately taken aback at the first meeting by the proposals that the ADC people laid on the table. In their own response to the ADC paper, they stated that they were 'surprised to find the district case for Highways increased over and above the original ADC submission'.[37] The ADC paper advocated that all highway functions, with the exception of strategic transport planning, should be directly conferred on the districts.[38] Moreover, this paper did not mention any limit on the size or nature of the district that could exercise such functions. In fact, it was stated that even in rural areas, where there was then no agency, a district should be able to take over the county's highway work.[39] Clearly, the ADC had changed their strategy with an eye to the General Election later that month and a possible change of government. The response of the ACC officers was to try to contain the discussion and insist that their remit was to consider only those changes that did not require legislation. So, a heated debate developed over whether a recent election statement by Michael Heseltine (then Shadow Secretary of State for the Environment) could or could not be interpreted as meaning that the Conservative Party would legislate in favour of the districts. The ADC representatives were of the opinion that it did, while the ACC representatives saw the task of the Working Party as simply to review the working of agencies and to make recommendations within the framework of the existing legislation.[40]

At the second meeting of the working party, the ACC officers put forward their reply to the ADC paper.[41] They found the ADC proposals totally unacceptable. It seemed to them that the ADC proposals would lead to increased costs, as the districts would have to establish their own enlarged highways departments and even create new ones. It would also mean the dismemberment of well-established centres of engineering expertise in the counties. The county officers questioned, too, the claims that the ADC made regarding the relative costs of agency and county work. The ACC had attempted to gather some information on these costs but had, in the time available, only been able to conclude that the evidence was less conclusive that the ADC paper had maintained. On the related question of the duplication of effort, of which the ADC had made a great issue, the paper argued that some duplication was the result of the districts themselves attempting to exercise policy making powers that belonged to the counties. Other duplication was often the result of the counties exercising their duty to supervise the work from the financial and the professional angles. The ACC attacked what was the basic claim of the ADC that the districts were more responsive than the counties. No evidence had been brought forward. Moreover, the county members were elected and had a legitimate claim to represent the local electorate. There was even a hint of technocracy in the ACC's claim that 'no evidence has been given that counties (within severe cash limits and lower budget provisions in real terms) have done other than assess priorities in the most logical way for maximum cost effectiveness'.[42]

The last major point in the paper was over transport planning. The ACC argued that the ADC had failed to grasp the nature of strategic transport planning. The ADC had assumed that, as in land use planning, strategy could be devised at one level and implemented at another. However, according to the ACC, transportation planning was about 'setting priorities and assessing programmes using public not private money - in other words, the proper allocation of fixed amounts of cash'.[43] In fact, over recent years, given financial restraint, the TPP process had become more a matter of 'fine tuning'. Further district involvement could only make this process more complex. In addition, the ACC argued that proposals could only increase central interference as the Secretary of State would initially become the final arbiter in cases of unresolved conflict.

It was quite clear, then, that the officers would not have been able to come to an agreement amongst themselves even if they had had the authority to do so, once the general election and then the change of government had thrown into doubt their previous assumptions about what was politically feasible. The only recourse was to settle down to define at least the areas of disagreement and so they produced a document setting out the various options, ranging from the extreme ADC position to the no-agency option of the ACC, and listing the advantages and disadvantages of each option from each association's viewpoint. The options they arrived at were as follows. Option five was the no-agency one, which the ACC favoured but had been reluctant to press. Option three was the solution for which the ADC had been pressing hard, and seemed to the ACC to be reluctant to abandon. Option four was just a modification of this position; while options two and one both involved the continuance of agencies. Option two, in placing a stronger duty on the counties to grant agencies to the districts, was compared with the compulsory sewerage agency,[44] and this was seen by the ACC officers as the ADC fallback position. Even so, the ADC were not happy with that option because it still preserved the agencies, though at least it created a presumption in favour of the districts.[45] The ACC objected to this option because it brought in the Secretary of State to decide disputes between authorities, adding up to more central interference in their view. Finally, option one left the counties with their existing responsibilities but proposed a further officer working party to consider ways of improving agency arrangements, if necessary, but without changes in the law. The main purpose of the working party would be to prepare a model agency agreement that the associations would then recommend to their members.

At this stage the membership of the County Surveyors Society was seriously worried that the new Conservative Government would favour the districts' case. The ADC appeared to them quite adamant that the bare minimum acceptable to them was that agencies should be made compulsory. The surveyors also thought that this was the thin end of the wedge that had, at its other end, the district acquisition of social services and education. They felt that their stance was eminently reasonable, involving as it did the possibility of some concessions towards the districts. The surveyors considered they were in the right and that the change the ADC was pressing could only worsen existing problems. However, they had little faith in the national politicians' judgement of the districts 'emotive case'.[46]

THE SECOND JOINT WORKING PARTY

The Association of Metropolitan Authorities had not remained unaffected by the controversy generated by 'Organic Change'. The controversy threatened to have very serious consequences for that Association as, unlike the ACC and the ADC, it included both district and county authorities. Obviously, in the metropolitan areas, the controversy was limited to the highways and planning fields and did not extend to education or social services which were already located at the district level. Nevertheless, there was a 'prospect of real confrontation between met county and met district over highways',[47] during 1979 as highway agency agreements came up for renegotiation. The AMA had already been under pressure from certain of the metropolitan districts which wanted to become highway authorities in their own right - the main districts concerned were Birmingham, Coventry, Manchester and Sheffield. It seems that the metropolitan county representatives had been meeting among themselves to form a common approach and these meetings strengthened their resolve to resist this challenge from the districts. The two Yorkshire counties took the most rigid stance on agencies. In West Yorkshire there was only one agency which one district, Bradford, had forced on the County through exercising its rights under Section 187 of the Local Government Act. Similarly, in South Yorkshire, there were four agencies in existence, to which Sheffield was trying to add a totally new agency. The West Midlands had a stormy history of negotiation over agencies since reorganisation but was still operating them at this time.[48]

The AMA had not been involved in the ACC/ADC Working Party, but were now brought into the discussions by the new Minister. The ADC and the ACC had met at elected member level to consider the report of the Working Party on highways, and the parallel one on planning, but they failed to reach any agreement on highways. On highway agencies it was clear that the ADC was not going to make any concessions until they knew whether or not the new government intended to legislate. At this point they intended to maintain their pressure on the new Conservative Government to legislate in their favour and certainly not to appear to be contemplating a compromise short of legislation. In contrast, the ACC strategy was to stress that legislation was unnecessary and that the problems with agencies could best be overcome by the counties taking over the work. But they also offered to compromise, on the basis of option one in the Report, by continuing some existing agencies.[49] Both sides, then, looked towards the incoming Conservative Government to break their deadlock.

The new Secretary of State for the Environment, Michael Heseltine, rejected the Labour Governments' 'Organic Change' proposals but was anxious to see the district-county disputes over planning and highways laid to rest. The Conservative Government did not rule out the possibility of legislation but made it clear that they would prefer to see a voluntary agreement. So, Tom King, the Local Government Minister, asked for proposals from the local authority associations by mid June. Having received these proposals, he called a meeting in July in the Department to discuss the issues both of planning and highways. The meeting was at member level, with four ministers in attendance, including Tom King and Kenneth Clarke representing the DTp, and elected members from the associations supported by officers from their secretariats. The proposal to set up a working party on the planning issue was agreed and once King made it clear that the government did not intend to legislate to give any districts highway responsibilities, the associations agreed to set up a working party as proposed in option one.

The new Working Party was to be at officer level and was to report within the next three months.[50] The Chairman was from the DTp, the under-secretary in charge of Local Transportation and Roads within the Department, but he was not a professional engineer. It is not unusual for civil servants to chair local authority joint association working parties on issues in which the government takes an interest. In these situations the central department is only involved on the clear understanding that the outcome is the responsibility of the associations, not of the department. A civil service chairman does have the advantage of being neutral in relation to the different local government interests and the arrangement also enables ministerial and departmental views to be fed into the discussions. Thus, in the case of this particular working party, the Chairman saw himself as 'only acting as a mediator or conciliator - though my preferred term is as a catalyst'.[51] The Chairman was supported by two officials from DTp who acted as the secretariat and five other senior civil servants, administrators and professional engineers.[52] It should be added that the civil service Chairman was already well known to the county surveyors but less so to the district engineers.

On the local authority side, each of the associations fielded six officer representatives.[53] As is usual practice with important working parties of this type, a pre-meeting was held at the Department in early August. This pre-meeting consisted of just the Chairman and three officers from the permanent staffs of the associations, who met to plan an agenda. Before the first Working Party meeting, the DTp circulated their own paper on 'Possible Principles on which Agency Agreements for Transport Functions should be Based'.[54] It suggested some principles on which agencies could be based and the scope of the agreements. It affirmed that strategic planning and resource allocation should remain with the counties, that agency arrangements must be in the interests of efficient operation but that the 'underlying presumption' should be that counties would be prepared to enter into agency agreements. Operationally, these agreements were seen as lasting for at least five years with the county unable to terminate them unless the districts were demonstrably inefficient; whilst the districts would enjoy maximum discretion, with the county just supervising to ensure value for money and adequate standards of work. The critical issue for the Working Party, as the Department saw it was, '...whether the Working Party should attempt to define, in general terms, the categories of districts which might be considered for particular agencies; this could be on the basis of population, dominance of the urban area or whether the district forms part of a wider urban area'.[55]

The first meeting of the Joint Working Party, in August 1979, was not much more than a general discussion of the issues. The ADC criticised the way in which the counties operated the agencies, maintaining that there was too much supervision both financially and professionally by engineers of equal status and claimed that the counties tended to interfere on purely local matters, especially traffic regulation orders. The county representatives, from the ACC and the AMA, stressed their financial and professional responsibilities. They argued that the districts could not expect a greater degree of freedom than the counties themselves had from the DTp in operating the trunk road agencies. This argument, which was inspired by the county surveyors, was to be a major plank in the county case. As one county surveyor observed, 'we thought that we could gain substantially by stressing that the Department of Transport must not be seen in an ambivalent light'.[56] The county representatives challenged the districts to produce evidence to support their complaints about agencies. The Chairman intervened here to stop the discussion being reduced to specific authorities, but he suggested that they should produce copies of what they saw as 'good' and 'bad' agreements from their different perspectives. The other outcome of the meeting was the establishment of two working groups, one on the technical aspects and one on the financial.[57]

Meanwhile, the professional engineering associations in local government were preparing their response to the ministerial statement and the DTp paper. The County Surveyors Society began to consider their response in one of their main committees and formed a working party, convened by the County Surveyor of Northamptonshire, which the two county surveyor members of the JWP attended. The paper this working party produced formed the basis of the final Code of Practice, according to one senior civil servant.[58] Their equivalent association in the districts, the Association of Chief Technical Officers (ACTO), was much slower off the mark. As one of their members commented, 'the County Surveyors upstaged us'.[59] Their response was drafted by their Secretary, the Borough Engineer of Cheltenham, an ADC representative on the JWP. A further professional group, the City Engineers' Group, produced a paper which was distributed to the local authority associations. It was, and is, quite unusual for this Group to submit joint views on a subject as, for most part, it is an information grouping with the objective of discussing matters of mutual interest among engineers in the larger cities, but clearly the preparation of their paper reflected their strong interest in the future of engineering at the district level.

The City Engineers' submission argued that the larger cities should be given agencies as of right. They proposed that the working of these agencies should be covered by a code of practice produced by the JWP. The reasons given for this were similar to the earlier ADC document - to reduce duplication, to provide services more cheaply, to allow the districts to deploy their staff more flexibly and to give a more understandable and democratic arrangement to the public.[60] There were two main principles on which they argued agencies should be based. The first was the need for a comprehensive approach in the big cities with their problems of urban regeneration, industrial development and so on. Thus, to their minds, 'the successful approach to these problems requires a single co-ordinated system which is best achieved by the district council exercising the maximum powers jointly as planning, housing and highway authority and as the initiator of development'.[61] The second principle was more specifically related to their particular professional claims. It was that 'district councils such as these large cities should be considered to be completely competent to carry out all the works referred to above and should be given discretion to do just this within the overall allocations';[62] an argument that implicitly questioned the existence of the counties, particularly the metropolitan counties. It is worth adding too that, in various supplementary points, the City Engineers' Group demanded further powers, they wanted the agencies in their areas to include county principal roads and to share in the preparation of the TPP. They did not envisage legislation but wanted, rather naively, to see the counties voluntarily giving these powers to the districts.

The debate between the counties and the districts, during the meetings of the JWP and its working parties, centred on three main points.[63] First, on the lines of the first paper produced by the DTp, they argued over whether or not the JWP should seek to produce recommendations relating to the types of district that would be permitted to acquire certain agency functions. Second, they discussed the duration of the agencies. Third, there was disagreement over the arbitration procedures in cases of dispute. Perhaps surprisingly there was little difficulty on the question of agency financial arrangements, particularly the fees that the districts charged to cover their overheads. This question was resolved quite quickly in the financial working group. The three other major points of contention were dealt with in the technical working party, which consisted of county and district engineers, with three staff members from the associations also in attendance.

The first point of contention was the major underlying issue. There were two aspects to this issue, the functions that districts of a certain type could operate and the feasibility of formulating different categories of district. On the first aspect the nature of the arguments deployed differed little from those fielded earlier after 'Organic Change' had been published. Of course, the intervention of the Minister had narrowed down the debate. The districts now gave up arguing against agencies per se, but still maintained that the counties should restrict themselves to a more strategic role and interfere less with the agency work carried out by the districts. They insisted too on their professional competence to carry out this work and presented county supervision as slight to this competence. Thus the district view was they they should be permitted to do the work subject only to their reporting back on the completion of the work and an overall budget. The DTp gave them some support over amenity maintenance, expressing the view that the districts could have more freedom on this. The counties' response was that there could be no possibility of the districts having more freedom in general and that the loosening of county supervision depended on the competence of particular districts.[64]

The county officers opposed the districts' arguments for the devolution of functions down to their level. They instead stressed the need for resource planning over a wide area and for an efficient and economical use of those resources. The county explanation of what might have seemed to a district to be excessive supervision was that the counties were attempting to maintain up to date information on local needs so that resources could be allocated according to overall priority. They stressed the importance of ensuring uniform and consistent standards across the individual county. They pointed out that this was precisely the way in which the DTp supervised the work done by the counties on trunk roads. Could the districts really expect less supervision than that the counties themselves had from the Department? This too was probably an argument directed towards the Department, in the debate over trunk road agencies, as well as towards the districts.

The districts were anxious to obtain agreement to a set of categories of district that would be considered able to exercise different levels of agency. The paper produced by the ADC, mentioned above, proposed such a set of categories. It delineated four types of district, the wholly urban, or nearly so, the mixed urban and rural with a centre predominating over the rural area, the mixed urban and rural just with a substantial urban centre or centres.[65] The first class of district would be able to exercise all non-strategic functions, the second of these functions just in the urban area, the third would have a limited agency arrangement and the fourth no agency arrangement (though the authorities in this category did not usually operate agencies in any case). However, this paper went on to define the metropolitan districts as cases of limited agency authority. This was in accordance with the 'Organic Change' White Paper which had argued that metropolitan districts should not become highway authorities because of the overriding need to ensure the application of uniform policies within the main conurbations. But the Chief Engineer and Surveyor of Manchester, who was representing the six 'core cities' in the metropolitan areas, could not accept that the metropolitan districts should be placed in a different category from the equivalent non-metropolitan districts.

A further objection to the ADC paper came from the counties, but their objection was to the whole principle of such categories. The county surveyors had already submitted their own draft code to the JWP, which had been drafted within their Society. This had contained a list of criteria against which the case for particular agency arrangements could be assessed by the county and the district concerned. It referred to criteria such as the extent to which the district already carried out functions, the districts' operational experience and the likely efficiency of the district. These criteria had been accepted by the district side early on in the JWP. Thus the county surveyors on the JWP felt that the districts were trying to pre-empt these earlier and agreed criteria. Their fear was that the districts would use categories of the type the ADC was advocating as powerful levers in local level negotiations. While the ADC engineers believed that unless the final code of practice had some 'teeth' in the sense of more specificity, then some counties could withdraw the agencies yet remain within the letter of the Code. The county surveyors feared that it would strengthen the district case at the local level and even encourage some districts to demand enhanced agency functions in accordance with their standing in such a list of categories. Moreover, the county surveyors on the JWP felt that they had already made enough concessions towards the districts in their own draft code and they were having doubts as to whether they could sell that compromise back to their elected members and their fellow county surveyors. However, the Chairman applied pressure to the county people, as one of them put it: 'the Department was quite clearly adhering strictly to the line taken by the Minister at the initial meeting and expecting the counties to permit agencies for highway functions where they can be performed efficiently and effectively'.[66]

Ultimately on this issue of the necessary criteria for the granting of agency, a compromise was reached, mainly on wording that ostensibly favoured the districts yet left it open for the counties to proceed as before. Reference to such criteria as urbanisation was made in the final Code but was not given the prominence or specificity that the ADC favoured.[67] It should be added that the final Code did not refer to any special approach in the Metropolitan areas, clearly a concession towards the metropolitan engineers' view.

Of the two other issues which were mentioned above, the duration of agencies and the arrangements for arbitration in case of disputes, neither was resolved within the time period originally specified by the Minister. At the second and the last meeting of the full JWP, the Chairman proposed that these issues be resolved through correspondence. So they dragged on after the formal conclusion of the Working Party and were resolved finally through the mediation of the three members of the associations' permanent staffs, who had sat on the JWP, and the Chairman.

With regard to the duration of the agencies, the original ADC position had been that the agencies should be granted on a permanent basis. This, it was argued, would give the districts a basic level of certainty on which to plan the deployment of their engineering resources, in particular recruit and develop professional engineering staffs. This demand was modified at the beginning of the JWP to a proposal that the agency agreements should last for ten years, with three years notice on both sides. This at least prevented the counties from unilaterally withdrawing agencies at short notice, something that the districts had complained about. The ACC agreed to this with the proviso that the period of notice was shortened to two years. Meanwhile, the AMA favoured a continuous agreement in the metropolitan counties, with three years notice and this again was incorporated in the final Code.

On the issue of the method of arbitration in the case of disputes between a county and a district, the ADC, in accordance with its earlier papers, argued that the Minister should act as the final arbiter. This was strongly opposed by the ACC and the AMA. Clearly the ADC felt that the Government still favoured the district side, and the county people too had little faith in the ability of the Minister to see their point of view. In the event, the conflict was resolved by the Minister himself. At the second meeting of the JWP, the Chairman said that the Minister was unlikely to want to act as the arbiter in such disputes;[68] instead, he suggested that the associations themselves should assume such a role. This proposal was not well received by the AMA and the ADC and there was no agreement on this issue by the end of the second JWP meeting.

The draft code, drafted by the DTp civil servants, was circulated to the associations for consideration in their committees. And an announcement was made in the House by the Minister, Tom King, in response to an inspired parliamentary question.[69] However, there was now to be a gap of five months before the final Code of Practice was published.

The delay in publishing the Code was the result of doubts about the Code and its possible consequences within the committees of the associations. The ADC went along with the code, reluctantly: 'In spite of their misgivings, the Committee finally decided, in the interests of unanimity in local government in responding to problems, that they would be prepared to accept the voluntary approach coupled with the proposed conciliation procedures'.[70]

In the same letter the ADC stated their serious doubts about the efficacy of a code lacking statutory backup powers allowing for ministerial intervention. They repeated their earlier commitment to the direct conferment of powers, or at least compulsory agencies. They made it clear that if the Code did not work, they would look to the Minister for some statutory backup. Similarly, the AMA Planning and Transportation Committee met and accepted the Code in the light of the Government's unwillingness to legislate, just having some doubts about the detailed wording of part of the Code. However, in the ACC, the issue proved more contentious. One elected member reported that they had never, in his recollection, had a vote in the Planning and Transportation Committee but in the debate over the Code, the Committee came as close to voting on an issue as he could recall.[71] The county surveyors who had sat on the JWP, had a hard job selling, what they had regarded as, the best possible compromise back to their Committee and, perhaps more importantly, back to their fellow county surveyors. As one of them commented, 'some colleagues felt we had been over-generous, the ACC felt that we had failed to achieve our objectives'.[72] There was strong pressure from some county surveyors, who had briefed their ACC representatives to oppose the adoption of the Code, these representatives then opposed the Code in the ACC Committees. The Transportation Committee Chairman saw much of the pressure against agencies and the Code as having come from the county surveyors through the briefing they gave to their members.[73] However, wider political considerations determined the issue at the end of the day. Firstly, it was clear that the ACC had to appear to be making some concessions towards the districts. Secondly, as a leading member of the Association concluded: 'It seemed better to accept this (ie the Code) than to argue publicly with a Conservative Government and to have public rows between Conservative districts and Conservative counties'.[74]

After further negotiations among the associations, with the Chairman of the JWP still acting as intermediary, over some minor points, mainly the conditions for the termination of agencies, the Code was finally published in April 1980 and circulated by the associations under their own imprint. The publication of the Code was announced in the House of Commons by Kenneth Clarke, the Parliamentary Under Secretary in the DTp, at the end of April:

> I am glad to say that the Association of County Councils, the Association of District Councils and the Association of Metropolitan Councils have all now agreed the Code of Practice drawn up by the joint working party established for this purpose. I understand that the associations are to publish the Code this week with the recommendation that their members should have regard to its provisions and the guidance it contains. The Government strongly commend the Code to individual local authorities and trust that it will provide the basis for county councils and district councils to work together, in a spirit of co-operation and understanding to provide an enhanced service to the public.[75]

AGENCY AGREEMENTS : THE IMPACT OF THE CODE OF PRACTICE

Before the impact of the Code is considered, the content will be outlined. The Code was of course exhortative and there was no way in which it could have been enforced on authorities. Its force was purely that of moral persuasion. It set out various considerations which districts and counties should bear in mind in negotiating and re-negotiating agreements. Thus, while it gave a list of criteria against which the case for a particular agreement could be assessed, they are by no means precise criteria of the type that the ADC had pressed for during the JWP. Nineteen points were actually listed and this alone suggests that the Code left considerable scope for local interpretation. Examples of the points included were the degree to which the district already carried out the functions, the ability of the district to carry out the functions and the size and extent of urbanisation in the district. The Code recommended that the agencies should last for ten years, with a period of notice of two years on either side, though this could be twelve months if one side considered that the other was in breach of the contract.[76] As already mentioned, there was provision for arbitration in the case of disputes or where the district felt that the county was unreasonably withholding the agency. In the case of disputes in a metropolitan area the AMA would set up a conciliation panel, while in non-metropolitan areas the ADC would first consider if they agreed with the district and, if they did, the ADC and ACC would set up a joint panel of one member and one officer each to attempt a conciliation. If this failed, the associations would then appoint an independent person to review the dispute, though his findings could not be in any way binding on the two parties.[77] Finally, the Code gave some administrative and financial guidelines on how the agencies should be operated.

It is difficult to give an overall assessment of the impact of the Code on the agency arrangements in the absence of a systematic survey. What it has failed to do is to halt the trend towards counties taking over agencies from the districts so that it has not even achieved the minimum that the districts had hoped. It definitely has not led to the establishment of new agencies. The counties, both on the elected member and the engineering side, have adamantly resisted any attempts by districts to create new agencies. In any case the delay in producing the Code meant that many agencies had been re-negotiated before the Code had been published. For example, this had happened in the West Midlands, where the Metropolitan County had come to a new agreement with all its districts except Birmingham by March 1979. And there were similar examples in the non-metropolitan areas.

As far as it has been possible to discover during this study, there have been only two cases in which the districts concerned have attempted to bring the arbitration procedures into play, though in both cases the counties have refused to go to arbitration. The two were Thamesdown in Wiltshire and the districts in the county of Clwyd. In the first, the negotiations reached an impasse early in 1981 in renegotiating the agency. Wiltshire claimed that Thamesdown, which is in effect Greater Swindon, demanded a hundred percent agency, which the County was not willing to accept. In turn, the District claimed that the County was not willing to negotiate a realistic agency.[78] The Chairman of the ACC Planning and Transportation Committee with ACC staff met representatives of the County at the ACC Headquarters, in accordance with the first stage of the conciliation procedure. However, Wiltshire refused to go to arbitration, despite the ACC's attempts to persuade them.

In the case of Clwyd, the County was in dispute with five of its six constituent districts. The County had proposed reducing the urban area in which the districts operated their agencies. When the districts refused to agree, the County then unilaterally withdrew the agencies, refusing the district request for an extension. The districts, led by Rhuddlan DC, approached the ADC who called for the conciliation procedures to be invokes. Rhuddlan was anxious to refer the issue to the Welsh Office but was held back by the ADC. From the ACC side, the Transport Committee Chairman and an ACC officer visited the County to discuss the issue and suggested that they went to arbitration, but again the County refused.

The two professional associations involved, the County Surveyors Society and the Association of Chief Technical Officers, were monitoring the implementation of the Code. The general consensus among them by the Summer of 1981 was that the Code had been, in the words of one county surveyor who had been involved in the negotiations, a 'dead letter'. It had neither led to the districts demanding new powers, as many county surveyors had feared, nor had it stopped the continuing erosion of district agencies by the counties.

CONCLUSION

Although disputes over agency agreements, in retrospect appear to be of little importance, at the time it was feared that they would be widespread and disruptive to the orderly running of local government.

Certainly, the engineers involved considered the issue to be important to the future of their profession. Local government reorganisation had built a new source of conflict into the profession as members of the profession disputed with each other who should be responsible for and who should perform highway engineering work.

However, it seems unlikely that the engineers could have kept the agency issue alive without the interest of the politicians. The district elected members, especially from the larger districts, used their influence in Parliament and the ADC to question the existing division of powers between county and district and to keep the issue on the agenda of both governments - in particular, using devolution as a peg on which to hang their case. In contrast, the elected members from the counties were anxious to preserve the status quo and resist attempts to place a further reorganisation on the political agenda. The county surveyors were in agreement with this line. They preferred to come to local arrangements with the districts rather than risk agencies becoming a national issue in a political climate which they read as being hostile to their position. While the district engineers feared that the county surveyors were simply picking them off one by one, and felt that agencies had to be made a matter of national level negotiations if this process of attrition was to be halted.

The Labour Government's 'Organic Change' proposals presented the districts with an opportunity to press for increased powers at the counties' expense. Although those proposals went some way to accepting the district case for more highway responsibilities, the arrival on the scene, soon after, of the new Conservative Government redefined the issue. The new Government saw the issue as almost entirely a local one that was not their responsibility. Their contribution was to knock heads together - providing departmental assistance to a Joint Working Party in the form of a civil service chairman and secretariat, and insisting that local government had to come up with its own resolution of the dispute. The remit of this new JWP was to arrive at mutually agreed and voluntary guidelines for agency arrangements.

Despite their misgivings the leading figures in the County Surveyors Society, including those surveyors sitting on the JWP, felt that their best strategy was to offer a compromise rather than risk failing to agree. Such a failure could induce the government to intervene, and Michael Heseltine, the then Secretary of State for the Environment, was considered by them to favour the districts. The Society moved rapidly. They produced their own draft code of practice and presented it to the JWP, to the surprise of the district officers. This draft set the agenda for the rest of the JWP's work. The district engineers replied with their own proposals, which were considerably more specific than those of the surveyors. They attempted to commit the county side to certain categories of districts which would be considered as having a prima facie case for running agencies. This idea was rejected by the county surveyors as it gave approval to existing agencies and appeared likely to encourage districts to demand more. But they made some concessions by agreeing to the inclusion of some, rather more vague, guidelines along the lines proposed by the district side.

The professional officers experienced problems in selling the final Code of Practice back to their associations. The county surveyors, in particular, faced the wrath of some of their colleagues who had briefed their elected members on the ACC to oppose the Code. The associations finally agreed to the Code because of the feeling that Conservative associations should not appear to be in conflict with each other and with a Conservative Government - an interesting case of the supremacy of community over policy.[79]

The agency issue is a good illustration of the ability of the county surveyors to act independently as an organised body and the political sophistication of their leading members. Throughout the course of the issue, they prepared themselves with arguments and kept the elected members well briefed about the county case. Their strategy was to attempt to define the issue as predominantly 'technical' in nature. In other words, they presented it in terms such as efficiency and economies of scale when it came to using expensive capital plant, and the need to have large authorities so that professional engineers could gain the necessary practical experience. In contrast, the district case was conducted in 'political' terms. It was argued that functions should be exercised by the level of government closest to the public, to ensure responsiveness and sensitivity to local needs. These arguments were drawn on by the district engineers, who experienced some difficulty in presenting their case in technical terms. Partly this was because they recognised that they had to mobilise the support of the district politicians and partly because they had difficulty in assembling a convincing technical case. In the absence of certain knowledge of the relative efficiency of county and district, the county surveyors could argue that the counties were more efficient simply on the grounds of their size. This keyed in well with the usual assumption that size is related to efficiency.

Unlike the county surveyors, the district engineers did not form a cohesive professional group. They had no umbrella organisation and were divided into different interest groupings which meant that it was difficult, if not impossible, for them to agree on a common line on agencies. As one city engineer observed in retrospect, 'the range of districts and the functions they administer militated against a concerted approach and the CSS took every advantage of this fact (in the course of the negotiations).'[80] The district engineers also suffered from not having a close relationship with the DTp civil servants of the sort enjoyed by the county surveyors. This reflects the district engineers' greater distance from the Department compared with the county surveyors. The latter, because of their small numbers, their involvement in trunk road administration and their single co-ordinating association had far more opportunity to form such a relationship with the civil servants.

This manifested itself, for example, in the informal meetings held three or four times a year, in addition to the usual round, between the three honorary officers of the CSS and the Under Secretary (Local Transportation and Roads), who was also the Chairman of the JWP, and sometimes the Deputy Secretary (Roads and Local Transport).

One former President of the CSS described these meetings and commented that they 'are very private and involve a full and frank exchange on issues of concern to both sides'.[81]

NOTES

[1] Royal Commission on Local Government, HMSO, June 1969 (Cmnd 4040), Vol 1, para.315.

[2] Ibid, para.315.

[3] Ibid, para.116.

[4] Department of the Environment, Report of the Committee on Highway Maintenance, HMSO, June 1970, p.76.

[5] It also provided, in Section 101, for districts to operate other services under agency agreements on behalf of county councils.

[6] The 'Big Nine' were Bristol, Derby, Hull, Leicester, Nottingham, Plymouth, Portsmouth, Southampton and Stoke on Trent.

[7] Royal Commission on the Constitution, HMSO, October 1973 (Cmnd 5450) (The Kilbrandon Report).

[8] Office of the Lord President of the Council, Devolution within the UK, HMSO, 1974.

[9] Office of the Lord President of the Council, Devolution - the English Dimension, A Consultative Document, HMSO, 1976.

[10] 'Memorandum of Observations on the Consultative Document: Devolution - the English Dimension', ADC, 1977.
At this time only 89 out of 296 non-metropolitan districts had no agencies, while 100 districts had full agencies. Appendix A of the 'Memorandum'.

[11] Ibid.

[12] This statement was made in response to a parliamentary question from the MP for Bristol North-east, Arthur Palmer. Bristol DC had been one of the most outspoken critics of the 1972 Act.

[13] Hansard, 25 July 1978.

[14] 'Most of the larger district councils are already involved to a greater or lesser degree in the provision of highway and traffic management services through agency agreements, although for a maximum efficiency these services need to be planned and co-ordinated over a wide area. However, we will consider the possibility that some districts could exercise these powers in their own right - especially with regard to traffic management - while preserving the counties' proper responsibilities for planning and allocating resources effectively. The precise distinction between districts who might seek some or all of these powers will continue to be studied.'
Hansard, 25 July 1978.

[15] This document was originally circulated to county surveyors in the autumn of 1978. See Report of Committee No 1, County Surveyors Society, January 1979.

[16] This may have reflected the strong views of the metropolitan county surveyor author.

[17] 'It is significant that within the experience of members of the Society, where there have been fairly full agency agreements which have been removed, the members are now settling down to their respective tasks in county and district, and there is very little of the friction which existed prior to the changes'. Report of Committee No 1, CSS, January 1979, para.6.1.10.

[18] Ibid, para.6.2.4 (original emphasis).

[19] 'They must also understand and recognise their position in the Principal/Agent situation as the executors of the Principal's instructions. This often leads to problems which can only be avoided by dedicated "professionalism" and the recognition of one's role'. Ibid, para.6.2.7.

[20] DoE Circular 131/72, 1972.

[21] The county surveyors interviewed stressed the importance they attached to this lobby in keeping the issue alive.

[22] It will be recalled from Chapter Four that this was a long-standing criticism of the DoE, DTp and the earlier Ministry of Transport.

[23] This emerged during interviews. As the present generation of county surveyors retires, it seems likely that the new county surveyors will look less to the Institution of Municipal Engineers, which will increasingly become a district based institution. Anticipating this development, there have been moves to bring the professional civil engineering associations closer together and, in 1983, the Institution of Civil Enginers and the Municipals amalgamated.

[24] In a personal interview.

[25] Organic Change, HMSO, January 1979 (Cmnd 7457).

[26] Ibid, para.4.

[27] Ibid, para.28.

[28] Ibid, para.29.

[29] Ibid, para.30.

[30] Ibid, para.34.

[31] Though, it should be noted, not in the provision of public transport where many districts remained important providers.

[32] ADC, 'White Paper on "Organic Change in Local Government" - Memorandum of Observations', March 1979, para.1.8.

[33] Ibid, para.3.34.

[34] Ibid, Appendix B. The ADC had submitted similar proposals in June 1978.

[35] The ADC also argued that the 'Big Nine' should have their own TTPs. The 'Big Nine' had proposed this previously in their own submission to the Secretary of State in June 1978.

[36] 'Note of a meeting of the Joint ACC/ADC Working Party on Highways and Traffic Management', held at Eaton House, 1 May 1979. The JWP included fourteen officer representatives of the two associations. From the ACC there were three county surveyors (Arthur Jacomb from Hampshire, Colonel Leech of North Yorkshire and Brian Oldridge of Cambridgeshire), one county clerk, one treasurer and a member of the ACC permanent staff. While from the ADC there were three district engineers, two chief executives, one treasurer and two members of the ADC permanent staff.

[37] 'ACC Paper on the Highway Function', May 1979. The earlier submission referred to is the 'ADC Response to Organic Change'.

[38] 'Highways and Traffic Management Functions', ADC, 25 April 1979.

[39] Ibid, para.16.

[40] 'Note of a meeting of the Joint ACC/ADC Working Party'.

[41] 'ACC Paper on the Highway Function'.

[42] Ibid, para.8.

[43] Ibid, para.16.

[44] Under the Water Act 1973, para.15.

[45] 'Report of the Joint ACC/ADC Working Party on Highways and Traffic Management', May 1979.

[46] The discussion in this paragraph is based on documents and interviews with those involved.

[47] Comment by permanent AMA staff member.

[48] See, John Mulvehill and George Morran, 'Inter-Authority Relationships in the West Midlands', Linkage Five, December 1980, pp.12-15.

[49] 'Dame Elizabeth has now written to Mr Tom King, Minister for Local Government and Environmental Services ... On highways Dame Elizabeth has emphasised that the Association pressed for Option 1 in Annex 1 if there is to be legislation; otherwise Option 5 is strongly pressed'. Letter from A C Hetherington,

Secretary of the ACC to all county chief executives, 13 June 1979.

[50] 'Note of a meeting with Mr Tom King at DoE', 12 July 1979. (Note written by a local authority association officer).

[51] In a personal interview.

[52] Including a regional controller and the Assistant Chief Highway Engineer (Road Maintenance).

[53] From the ACC there were two county surveyors, one chief executive, one county secretary and one treasurer plus a permanent ACC staff officer. While from the ADC there were two district engineers, a treasurer, a chief executive and two ADC officers. Finally, the AMA sent along two county surveyors, two district engineers, a county chief executive and a district chief executive plus an AMA officer.

[54] 'Possible Principles on which Agency Agreements for Transport Functions should be based', DTp, 3 August 1979.

[55] Ibid.

[56] County surveyor in personal interview.

[57] Paragraph based on the 'Minutes of the First Meeting of the Joint Working Party', DTp, 20 August 1979.

[58] Personal interview with civil servant.

[59] In a personal interview.

[60] City Engineers Group, 'Review of the Local Government Act, 1972, Highways and Traffic Management', 1979.

[61] Ibid.

[62] Ibid.

[63] The following discussion is based on the minutes of the meetings and interviews with the participants.

[64] 'Minutes of the Meeting of the Technical Working Group of the JWP', 10 September 1979.

[65] 'Suggested Guidelines for Categorisation of Districts', ADC, September 1979.

[66] In a personal interview.

[67] ACC, ADC and AMA, 'Highway and Traffic Management Agency Agreements: Code of Practice', April 1980, para.4.2.

[68] Minutes of the Second Meeting of the JWP, DTp, 4 October 1979.

[69] 'I understand that this Working Party has now prepared a draft

Code and that it is at present being considered by the associations'. Hansard, 23 November 1979.

[70] Letter to the Chairman of the JWP from the Secretary of ADC, 23 November 1979.

[71] In a personal interview.

[72] In a personal interview.

[73] Interview with the Chairman of the ACC Planning and Transportation Committee.

[74] Ibid.

[75] Hansard, 29 April 1980.

[76] ACC, ADC and AMA, 'Highway and Traffic Management Agency Agreements', para.5.4.

[77] Ibid, Appendix 4.

[78] Municipal Engineering 10 February 1981.

[79] See chapter two.

[80] In a letter to the author.

[81] In a personal interview.

7 Highway maintenance

INTRODUCTION

The issue to be examined in this chapter differs in several ways from that considered in the previous chapter. Unlike agency agreements, the professional community did not become split over the issue of highway maintenance but rather was united by it. The debate was also dominated by the professionals who were actively attempting to persuade their political masters, at both levels of government, of the need to increase spending on the maintenance of the roads. This chapter describes the professionals' definition of the maintenance problem, which involved attempts to define the problem in technical terms and, therefore, to move it away from the political arena. The account will begin by considering the Marshall Report, which was the first major central government initiative in the field and raised the possibility of national standards for the condition of roads. Then, the significance of post-Marshall developments will be discussed - in particular, the development of new, sophisticated methods of assessing the structural condition of the roads and the establishment of an annual Road Maintenance Condition Survey, to provide 'objective evidence of the state of the roads' - a Survey that became an issue in central-local government relations.

THE ADMINISTRATIVE AND TECHNICAL BACKGROUND

Before the issue of highway maintenance is discussed in depth, the administrative and technical background will be outlined. For administrative purposes, roads are divided into two types: motorways and trunk roads, and then other roads. Responsibility for the latter falls to the counties, except in London where responsibility is divided between the GLC and the London Boroughs; while for the former type of road the Department of Transport is the highway authority. However, the counties perform various maintenance tasks for the DTp roads, under agency agreements, according to which the costs of any agreed work on DTp roads are reimbursed on an annual basis plus a percentage of those costs to cover administrative overheads. This percentage is negotiated at the national level through the ACC. The DTp itself allocates funds to the regional offices, and the regional controllers (transportation and roads), previously the divisional road engineers, have considerable discretion in allocating those funds among the counties in their region. Since the mid nineteen-seventies, this method of allocation has become more systemised with the introduction of a road condition assessment system which the Department has required all agency authorities to use on their trunk roads.

However, the counties remain the highway authorities for the great majority of roads - 94% of the total mileage, though in terms of vehicle usage, it is rather less. Until local government reorganisation the maintenance of non-trunk roads was funded through the local rates and the rate support grant from central government. The latter was based on a formula such that central government had no control over expenditure on particular services. But in 1974/5 this changed with the introduction of the transport supplementary grant.[1] Since then, the government has been able to take a closer interest in local roads, and the significance of this change will be discussed later in this chapter.

On the technical side, different levels of usage mean that the problems of maintenance are not the same for all roads. Those major roads that carry large volumes of heavy commercial vehicles, which have by far the most damaging effect on roads, present quite different problems of maintenance from less heavily trafficked roads. Motorways, in particular, carry about 25% of all heavy goods vehicle mileage, although they only represent about 1% of the total road mileage.[2] The paraphernalia associated with motorways increases the problems of maintenance, signpost gantries, emergency lighting and telephones, crash barriers and so on, all have to be maintained, but it is the difficulties of keeping the road surface together under the onslaught of fast moving heavy vehicles which present the most serious problems. The special treatments required on these high speed roads involve expensive capital equipment which is provided by the counties. For trunk and motorways the Department specifies certain standards with respect to materials, methods of construction and design objectives. These standards are contained in several manuals issued by the DTp and are regularly updated to follow new developments. It is important to stress that these are specification standards and are quite different from the general road condition standards to be discussed in this chapter. However, through these specifications, the Department is also able to exercise a strong influence over the maintenance practices of counties on their own roads, as the specifications provide a well-established and tested, a ready made guide for the counties which the Department encourages them to use.[3] On the other hand, extensive consultation takes place with the county surveyors over these standard specifications before they are issued, so the latter have considerable influence over the formulation of new standards. There is a considerable amount of co-operation between the DTp and the County Surveyors Society which participates in joint working parties on engineering matters with the Department.[4] So it is broadly true to say that the DTp is implementing an engineering consensus rather than imposing a set of predetermined standards.

The next technical point that should be explained is the nature of the different maintenance tasks. Maintenance is usually separated into four categories: winter maintenance, routine or cyclical, aids to movement and safety, and structural maintenance. Winter maintenance is self-explanatory. Routine maintenance refers to those tasks which are performed on a regular basis, such as grass-cutting and road sweeping, and is sometimes called amenity maintenance. Maintenance falling under the heading of aids to movement and safety, covers work such as repair of road signs. Finally, structural maintenance is closest to the traditional engineering tasks and to the highway engineers' own interests. It refers to work on the surface and the structure of the road, the road edge and and the footway. The biggest element in structural maintenance is related to work on bituminous carriageways. As background to the later discussion in this chapter, which is predominantly concerned with structural maintenance, it will be helpful to distinguish between what the engineer regards as short term and as long term maintenance treatments on the carriageways. Short term treatments include patching up cracks or holes in the road and surface dressing. Surface dressing refers to the laying of a thin surface of loose chippings and tar or bitumen binding which is intended to seal the road against further erosion through water seepage which would otherwise weaken its foundations. It also improves the skid resistance but not necessarily the rideability of the road for the user. However, according to the engineers, this type of treatment will leave underlying structural defects in the road unaffected. In time the surface of the road will collapse unless these defects are remedied. This is done through long term treatments such as resurfacing, strengthening by overlay or by reconstructing the road. Of course these treatments are considerably more expensive than the short term measures. The difficult problem facing the engineer is at what point these different treatments should be used in the life of a road, what is termed 'fixing the intervention level'. Various road condition assessment systems are intended as guides to fixing these intervention levels.

THE MARSHALL COMMITTEE ON HIGHWAY MAINTENANCE

The Marshall Committee was set up in response to Treasury pressure on departments to review public expenditure at the time of the Labour Government's 1968 expenditure cuts. Attention was probably drawn to maintenance because it was an area in which those at the centre only had a vague idea of how money was being spent out in the localities by what must have seemed an administrative morass of over 1,000 authorities with some highway maintenance responsibilities.[5] At about the same time, the County Surveyors Society and the Road Research Laboratory had also set up joint working parties on maintenance, so it was also a concern of local authority engineers. It is clear from the Committee's remit that those at the centre were interested in financial control and the possibility, or perhaps the necessity, of uniform standards across the country.[6] But is is not known whether those at the centre saw the Committee as a precursor to further central control. Certainly there was a strong body of opinion amongst the county surveyors which regarded the Ministry's new found interest in the issue with suspicion. They regarded maintenance as a last 'bastion of local autonomy',[7] and as a traditional function of highway authorities; a view strengthened by the fact that most of the county surveyors would have, at this time, come up through maintenance rather than design work, as the establishment of large in-house design teams in the counties had been a recent development. The Committee was also being set up at the same time as negotiations leading up to the creation of the road constructions units, which were again seen by many surveyors as a threat to local autonomy. Those engineers opposed to the Committee felt that it and the RCUs were part of a concerted attack on local autonomy and, therefore, on their own professional freedom. In contrast to this view others, generally the younger surveyors, began to discern new professional possibilities in the government's interest in roads. It was these who later played a part in spreading the new gospel according to Marshall.

The Marshall Report emphasised the wide variations in maintenance expenditure among highway authorities. The Road Research Laboratory carried out a survey which revealed startling differences in expenditure on both county roads and trunk roads; for example, on trunk roads maintenance expenditure varied from less than £800 per mile per annum to over £5,500 among the counties.[8] The Committee believed that only about half of these variations could be explained in terms of variation in weather conditions, traffic flows and road widths. They concluded that:

> These figures strongly suggest misapplication of resources. ... Our failure to explain more than about half the variation has confirmed the large part played by subjective judgements in producing the present differences between authorities.[9]

They regarded these differences as the result largely of the 'interplay of (engineers') judgements and the views of council members'.[10] Further, subsidiary reasons for these differences were found too in what was considered the failure of many authorities to organise their maintenance on up to date, efficient lines.[11] So, the object of the Report became to move the decision process over maintenance in the direction of a 'detached and rational appraisal of needs'.[12] Essentially, this meant instituting a set of output measures or standards in highway maintenance which would remove from 'the field of judgement as many aspects of the subject as are capable of measurement'.[13] This approach was very much in accord with the then current interest in local government management circles, in new techniques like PPBS (planning, programming and budgeting system), output budgeting and corporate management. Hedley Marshall, himself one of the leading authorities on local government finance,[14] persuaded the, initially, rather sceptical engineers on his Committee of the practicality of such standards and their value to the engineer back in his authority:

> What we were after as a Committee, was setting up a method at the local level by which engineers could say to their authorities, on the objective data, that you should spend 'x' amount on the roads. Such criteria do not exist. But you have to have some kind of criteria, the Committee tried to define these criteria. With these criteria the engineer is in a stronger position at budget time, and well in advance of the other local authority services.[15]

Another aspect of the Committee's interest in standards was concern with the structural condition of the roads. They had been impressed by the claims of the County Surveyors Society, in their evidence to the Committee, that 2,000 miles of road required major reconstruction as the result of inadequate maintenance and that a further 6,000 miles were in danger without improved maintenance.[16] The Ministry too had recently had to undertake several emergency works on trunk roads, which again had drawn attention to the importance of maintenance.

Finally, there was also an important central-local government relations aspect to maintenance standards. One concern was that highway authorities should improve their own maintenance practices in line with national standards.[17] The other concern was that decisions on trunk road maintenance were being taken in the Ministry on the basis of guesswork, looking at what was spent in previous years or relying on a standard distribution of funds per mile. The Committee saw the introduction of standards as enabling these decisions to be made on a more informed and territorially equitable basis, and as a means of improving the upward flow of information to the Ministry. Incidentally, the Committee also criticised what they considered to be the excessively detailed control over agent authorities exercised by the divisional road engineers. They recommended that the DREs should limit themselves to overall policy and financial control.

Despite this criticism of the Ministry, the Report stressed the importance of the Ministry taking the lead in ensuring that the Report recommendations were implemented. The view was that there was a 'risk, indeed a certainty, that development will be piecemeal and ineffective unless there is some central effort to ensure that changes are common to all authorities and adopted over the same time scale'.[18] A major way of ensuring this was for the Ministry to introduce the new standards on trunk roads as the agent authorities were 'unlikely to retain different methods of management on their own roads from those which they use as agent authorities'.[19] The other main channel, through which the new standards and approach to maintenance were to be encouraged, was the creation of a joint central-local government committee. This standing committee would, according to the Report, keep in touch with new developments, encourage research and ensure that authorities received advice when necessary.

The Marshall Report set the agenda on highway maintenance for at least the next ten years. As one county surveyor commented, 'Marshall put maintenance on the map',[20] and as a civil service engineer too has commented, 'road maintenance emerged on the national scene after long years of piecemeal local determination'.[21] It introduced the idea of national standards into the maintenance field, a change that was strongly favoured by all sides at this time.[22] The local authority engineers, for the most part, saw the idea of standards as introducing an important new argument in their case for increased expenditure on highway maintenance, both with central government and their elected members. At the same time it chimed in well with the central interest in improving the flow of information upwards, particularly in giving the centre a 'value for money' guide. Undoubtedly, too, in the Ministry, standards were seen as an important move towards uniformity in a field where local variations were difficult to justify. The rest of this chapter will be concerned with the further development of national standards in maintenance.

THE MARSHALL REPORT STEERING COMMITTEE

The Report was not ignored by the newly created Department of the Environment, which incorporated the former Ministry of Transport.[23] In 1970 the new Conservative Minister for Local Government, Graham Page, held a news conference to launch the Report and announced the Government's intention to implement it. Accordingly a committee at officer level was appointed, in line with the Marshall recommendations, consisting of departmental officials and local authority association representatives. The Secretary of the County Surveyors Society, who had been a member of the Marshall Committee, was one of the members of this new Marshall Report Steering Committee. He, and the other county surveyor on Marshall, had already begun touring the country speaking at meetings of the regional groups of the Institution of Highway Engineers and with elected members and engineers in individual counties. Their intention was to sell the ideas in the Report and to win round those engineers sceptical of its value.[24]

The Marshall Report Steering Committee began work quite quickly under the chairmanship of the Ministry Chief Highway Engineer. It directed its attention first to the maintenance of trunk roads and recommended that the initial standards contained in Marshall should be applied to trunk roads 'as a means of measuring the condition of the nation's roads'.[25] Some further work was done by the Department on the costings involved and in November 1971 the Minister announced that work would be done to bring trunk roads up to Marshall's initial structural standards over the next two years.[26] So the Report had already begun to show its value to the Departmental roads men in their competition over resources. As one prominent county surveyor noted, the Report 'was responsible for arresting in many counties what was regarded by most practising engineers as a rapid decay in trunk road maintenance'.[27]

This two year programme meant that the expenditure on trunk road maintenance doubled over the two years.[28] But the hope of the Committee had been that, once this major programme had been completed, funds would then be redirected towards county roads, particularly principal roads.[29] In the event these hopes were dashed. In July 1975 Dr Gilbert, the then Minister of Transport, announced future cuts in spending on trunk road maintenance and urged local authorities to follow his lead on their own road maintenance budgets. This reduced hopes of more money becoming available in the near future.

Before the response of the county surveyors to this development is described, the work of the Marshall Steering Committee will be examined. Under the umbrella of this Committee four working parties were created. The first two were really a resumption of earlier working parties that had been set up jointly between the County Surveyors Society and the Road Research Laboratory.[30] These two working parties had been suspended once the decision to set up Marshall had become known, as there seemed no point in both covering the same territory. Though the fact that they were established in the first place was clear evidence of the growing interest in maintenance in both official and professional circles. The two other working parties were set up in parallel and their membership was drawn from the urban highway authorities. Two of the working parties, a county and an urban, dealt with ways of improving the organisation of maintenance and the methods and procedures used in order to maximise the use of available resources. The two other working parties were looking at the application of existing and the future development of machines, materials and methods. They intended to focus on the establishment of 'objective standards of maintenance based on the economic balance of the costs and benefits of carrying out maintenance'.[31] Each working party consisted of six or seven local government engineers, both from chief officer and assistant chief officer levels, together with representatives from the Road Research Laboratory and DoE, but they were chaired by local government engineers. The county ones were chaired by the West Sussex Surveyor and Secretary of the CSS (the technical group), and by the County Surveyor of Derbyshire (the organisation group). The main method of working was through encouraging and sponsoring experiments in individual local authorities. The 'emphasis is on active participation by the local authorities, particularly in providing data and opportunities for controlled trials of improved methods and techniques of maintenance'.[32] So the groups were to act as catalysts and as disseminators of successful practice. Finally, it is interesting to note to which bodies they were officially working. The RRL Annual Report stated that: 'The working parties are under the general guidance of the Laboratory's Research Committee on Design and Construction and the County Surveyors Society's Committee on Maintenance, to which they are required to report progress'.[33]

The two major initiatives with which the working parties became concerned were, to give them their acronyms, CHART and RATE. CHART was and is a system for arriving at assessments of the condition of the road surface which then can be used as a basis for determining maintenance priorities; RATE, or Resource Allocation and Techniques of Estimating, was and is a computer-based system intended to supply information in the estimation of the cost of work.

CHART was based on work begun in the West Riding CC and subsequently in its successor, South Yorkshire. In essence, it was a system for collecting data about the condition of road surfaces and assembling it so as to arrive at an estimate of the relative conditions of different stretches of road. It is important to note that it was based on a visual inspection, at least in its initial applications. Quite early on there was some professional dissatisfaction with this visual basis. In CHART, the need for structural maintenance was largely indicated by visual measurements of wheel-track rutting and cracking, which were not regarded by engineers as reliable indicators of structural failure. For example, the movement of unstable surface materials as well as structural failure can cause wheel-track rutting.[34] This led the Transport and Road Research Laboratory to work on a new type of machine, invented in France, the deflectograph. This machine measures the degree to which the road surface gives, that is 'deflects', when a heavy axle passes over the road at creep speed. The measurements on site are then combined with a count of the heavy lorry traffic passing over the road under examination, to arrive at an estimate of its residual life; in other words, the number of years before strengthening is likely to be needed.[35]

CHART took considerably longer to develop than was originally thought. In fact, in its fully developed form, with various computer programmes and a hand held micro-processor, it has only been available since 1980, however it was already used on trunk roads in its earlier provisional form. The delay has meant that many authorities have gone ahead with their own systems in the interim. Apart from the delay, another major reason generally given by engineers for developing their own system, has been the need to produce something less expensive. CHART involves a hundred percent sample of roads in a category, thus for the DTp all the trunk roads in the country are CHARTed. To do this is feasible, since the trunk road mileage of a county is quite small in relation to its total mileage and, in any case, the cost is covered by the DTp. But even to do such a survey on the principal roads in a county can prove expensive as their mileage is several times greater than that of trunk roads. So, during a period of falling resources for roads, it was difficult to justify such expenditure when there was little chance of undertaking all the remedial work uncovered. A number of counties, then, have developed systems which replace the hundred percent sample with a smaller sample, although most of these systems are still based on CHART principles. A further reason for the adoption by counties of diverse systems has been the problem of making them compatible with the counties' computer installations, which would not have been selected with the needs of highway maintenance uppermost.[36]

Despite the problems with CHART, the Transport and Road Research Laboratory has worked hard to persuade the counties to adopt the system. Of course, this represented part of DTp's continuing interest, since Marshall, in improving the maintenance practices of local authorities and the quality of the information passed on to the Department. Seminars on CHART were held at the TRRL and the staff there wrote leaflets and articles in the engineering press as well as addressed local and national meetings of the Institution of Highway Engineers, to which many of the TRRL staff belong, and other professional meetings. A CHART Users Group was formed to pool experience and to feed it back to the TRRL staff developing it. The Group was chaired by the Director of Transportation at the GLC - the GLC is a major user of CHART and had been one of the first authorities to adopt the system.[37]

There were further, political, reasons why many county surveyors were interested in CHART. The first has to be seen against a background of growing concern among county engineers that they were losing ground in terms of the deference given to their professional judgement by elected members - as one said: 'at one time the county surveyor's words were those of a god, now that's no longer true'. Consequently, their recommendations were being increasingly questioned by elected members anxious to press for maintenance schemes in their own wards, with shrinking highway budgets. Against this, systems like CHART offered engineers a defence to back up their professional judgement. The potential of CHART lay in how it could be used to persuade members to restrict themselves to taking generic decisions on allocating funds among groups of schemes and not to interfere with decisions on particular schemes. The second reason relates to agency arrangements with the districts. One of the major problems with these arrangements was satisfying such districts that they were getting a fair share of maintenance funds compared with districts in the rest of the county. The use of an assessment system could enable the county to argue that these funds had been allocated according to need as measured 'objectively'.[38]

The second development which was progressing during the same period as CHART, was RATE. Work on this costing system was done under the umbrella of the joint County Surveyors and TRRL Road Maintenance Working Party, as was CHART. RATE had originally been developed in the former Lindsey County Council and it had come to the notice of the Marshall Committee, which had recommended that further work be done on the system. The 1974 TRRL Report describes the system as enabling 'the engineer to produce, for any given job of maintenance or construction, accurate estimates, based on work measurement and method study data, of the costs and resource requirements related to the bill-of-quantity items'.[39] So whereas CHART related to the condition of the road surfaces and helped determine priorities, RATE assisted in estimating the cost of the proposed work. Clearly, in the planning of work, it is important to know the likely costs of maintenance jobs.

Developmental work on RATE continued in the successor to Lindsey, Lincolnshire County Council, under a research contract with the TRRL.[40] Later, a trial installation of the computer programme was tried out in Cornwall's highways department, arranged through the Surveyor who was a member of the Maintenance Working Party. This appears to be a quite usual way for TRRL to set up experiments in authorities, using contacts on their joint working parties. However, the final system was not unveiled until October 1979 at a 'symposium' at the TRRL attended by 80 engineers,[41] though it had been originally planned to take place two years earlier. The system had already begun to be applied in some authorities after the initial experiments, and already a RATE User's Group had been formed to exchange information among those already employing RATE. Clearly some importance was attributed to RATE by the DTp as soon after the seminar the Standing Committee on Highway Maintenance established a RATE Sub-committee. The Sub-committee was under the chairmanship of a county surveyor and comprised local authority officers and officials from the Highways Engineering branch of the DTp and TRRL.[42]

THE STANDING COMMITTEE ON HIGHWAY MAINTENANCE AND THE ROAD MAINTENANCE CONDITION SURVEY

In September 1974 DoE proposed that it and the local authority associations should set up, at officer level, a joint committee on maintenance as a successor to the Marshall Report Steering Committee. The associations responded favourably and the first meeting took place in July 1975.[43] The new committee, the Standing Committee on Highway Maintenance, comprised civil service engineers, administrators and statisticians as well as local government engineers and treasurers. The chairman was a civil service engineer, the Director of Highways (Planning and Management), until 1978 when he retired and was replaced as chairman by an administrative civil servant; another indication of the loss of power by the professional engineers within the Department.

The Department was concerned with improving the efficiency of maintenance so it may not have been entirely a coincidence that the Committee was formed at a time when the Government was contemplating cuts in highway maintenance expenditure. These cuts were announced within days of the first meeting of the new Standing Committee on Highway Maintenance (SCHM). Dr Gilbert, the Minister of Transport, announced the Government's future intention to reduce expenditure on trunk road maintenance by 15-20% and went on to encourage local authorities to follow suit.[44]

Soon after the ministerial announcement, the County Surveyors Society carried out a survey on road maintenance expenditure in July 1975. They claimed that no non-metropolitan county had been able to keep up with inflation in its maintenance budget. So the deterioration in the road network was approaching a critical point, and they estimated that 47% of the total mileage of principal roads would need virtual reconstruction within the next ten years.[45] Their message was quite clear. In their view, unless there was a continuing programme of remedial work, a crisis in the condition of the roads would arise, involving widespread road failures. These failures would be both expensive to repair and costly in terms of traffic disruption, since the repair works would mean rebuilding parts of the roads affected or, at the very least, extensive re-surfacing.

The threat of reductions in maintenance expenditure added new impetus to an idea that had already been discussed in county surveyor circles. The idea was for a national road maintenance condition survey that would give 'objective' support to the surveyors' claims that the condition of the roads was worsening. As one member of the SCHM commented, with reference to the Minister's announcement: 'A better catalyst for redoubling efforts to develop an objective system for measuring the condition of roads would be difficult to find'.[46]

The idea of an annual survey had already been mooted in an ACC working party on the condition of minor rural roads that had been set up early in 1975.[47] The working party was the result of pressure from the county surveyors who took the view that minor county roads were deteriorating beyond acceptable levels. This view was reinforced by the fact that many local authorities had heeded the Marshall Report's recommendation that counties should redistribute funds towards their principal roads and away from minor roads. The Marshall Report had argued that resources should be concentrated on the more heavily trafficked roads and criticised what it saw then as excessive spending on the less important minor roads. But the ACC Report stressed that there had been changes in the pattern of road usage, in particular the use of heavier lorries to deliver materials to farms and to collect farm produce. These changes were seen as having a serious impact on the state of country roads.

The county surveyors were well represented on this working party. There were five surveyors out of a membership of thirteen, the other members being one treasurer, one TRRL scientist, two National Union of Farmers representatives, and a representative each from the Freight Transport Association, the AA and RAC, DoE and the Welsh Office. It was serviced from DoE with a DoE officer acting as the Secretary to the group. This type of servicing arrangement is not unusual. The ACC lacks the staff to undertake this type of work and if the central department is interested in the issue it may help by providing support staff. However, this is done on the clear understanding that the final reports are issued on the authority solely of the association or associations concerned. The advantages of this arrangement for both sides are obvious. The associations get extra support while the department can encourage or even defuse local government interest in a policy area yet not themselves become committed to any statements which can be used against them later.

The working party accepted the surveyors' view that the condition of rural roads was deteriorating. This fact was attributed to the falling maintenance expenditure and to the increasing use of large lorries damaging verges, road edges, drainage and footways. However, rather optimistically, but not surprisingly given the set of interests represented, it concluded that the extra damage caused by larger lorries was more than offset by the savings the community made through the use of the lorries. Consequently, heavier lorries, the Report argued, should be accommodated by road improvements, in particular, by counties distinguishing certain key distributor roads which would be selectively maintained and improved at 'some additional cost to cope with modern traffic needs'.[48] This was very much an engineers' response to the heavy lorry problem, 'give us the resources and we can cope'.[49] The other major recommendations of the Report were that there should be a scheme for monitoring the state of the minor road network, and that the newly formed Standing Committee on Highway Maintenance should be asked to sponsor such a scheme. This appears to be the first mention of such a survey in a public document. The heavy representation of county surveyors on the working party means they must have supported the idea - and probably sponsored it. The County Surveyors Society would certainly have discussed a proposal of this type in one of its committees before five of the Society's members would have gone along with it. In fact, it seems highly likely that the idea came from the Society in the first place.[50]

Early in 1976 the Standing Committee on Highway Maintenance discussed the proposal for such a survey and they finally agreed to one covering all classes of roads, not just minor roads, which would give an overall picture of the state of the nation's roads. The county surveyors were the people who had to persuade their elected members to go along with the proposal, as it required the counties to perform the actual collection of data which was seen as inevitably involving them in extra expenditure. In the event, the survey had 100% co-operation from the counties, something which the committee regarded as an important achievement. This degree of support and co-operation from the county surveyors reflected their strong conviction that the state of their roads was worsening and that a survey of the type proposed would demonstrate beyond doubt that serious deterioration was taking place. What many of them thought was needed was an 'objective survey' to back up their professional judgement. As long as they based their arguments for more funds or just for retaining existing levels of expenditure, only on professional judgement, they were always liable to be seen as trying to protect their own empires. In other words, they did not want to be seen as pushing their own sectional interests. So it was widely hoped by the county surveyors that the results of such a survey would support their case for funds, vis-a-vis both the government and their own elected members.[51] The surprise, and in some cases anger, that many surveyors expressed later when the results failed to provide the evidence desired, indicated the high expectations they had of the National Road Maintenance Condition Survey.

Once the County Surveyors Society had agreed to it, the DTp prepared a paper on the form of a possible survey. But it was emphasised by the Department that the final decision on the nature of the survey was the local government engineers', according to one civil servant: 'they had been told that it was their own show'.[52] The SCHM went ahead and formed an Executive Group to develop and organise the National Survey. The group reflected the same type of membership as the main committee but was chaired by a county surveyor.[53] So the first experimental survey was conducted during the autumn of 1976. The group was able to get the survey off the ground this quickly because of the interest among the county surveyors and because it was based on

CHART, which was already used by all the counties on trunk roads. The object of the survey was to monitor the effects of changes in maintenance expenditure over a period of several years - so one year's results were of little value unless they could be compared with the results for other years. Thus the aim of the NRCMS was to uncover the overall trend in the condition of the country's roads.[54] Information that could then be used by central government to make decisions with confidence.[55]

In designing the Survey, then, it was considered vitally important that the results would be generally regarded by all sides as reliable and as 'above suspicion'.[56] Thus the lengths of road were randomly sampled by DTp statisticians who told the county staff, who actually conducted the survey, which stretches of road to examine. While to avoid any criticism that these sites could be deliberately neglected, after the first three surveys one half of the sites were changed annually. The Survey collected data relating to ten different types of defect on the road and four more to the adjacent kerb and footway. Not all of these types of defect would be apparent to the ordinary driver. In fact, some defects which the driver might regard as serious in that the quality of his ride is affected, are not judged as serious by professional engineers. For instance, irregularities in the surface of the road will affect the ride but will be of less significance to the engineer. Moreover, some of the more serious defects in the engineer's view go unnoticed by the driver. For example, minor cracks in the wheel tracks of a highly trafficked road are generally seen by engineers as symptomatic of a deep seated weakness, which will ultimately lead to a rapid deterioration of the structure of the road unless remedied. The professional analysis is that these cracks allow water to penetrate under the surface, causing erosion by leaching and freezing. This sort of structural failure or potential failure is often not obvious to the layman. The Survey, then, sought to measure defects such as the extent of wheel-track cracking, potholes, edge deterioration, footway deterioration and kerb deterioration.

The Survey was based on a visual inspection of the roads. The Committee did contemplate the use of some of the scientific aids that had been developed, but decided against using them, at least in the first instance. They were seen as having several important disadvantages. The deflectograph, which was mentioned earlier, was regarded as yielding more reliable results than a visual inspection and was strongly favoured by the TRRL. But it was an expensive machine and was mainly restricted to trunk roads at the time of the first survey. Another machine, SCRIM, which measured skid resistance, was again expensive both to acquire and to use, whilst many engineers regarded it as obsolete. Finally, the bump integrator, which is a means of measuring the rideability of a road surface, was ruled out as it can only provide valid results over longer road lengths than were practicable in the Survey. The further reason why machines were not used in the early surveys was the desire of the SCHM to secure 100% co-operation from the counties.[57] To have specified the use of such expensive machines could have deterred counties from participating in the survey. However, this decision proved controversial later when the results of the Survey were released.

In the meantime, before the results were released, the engineers kept up their pressure on the maintenance issue. In 1978 the County Surveyors Society published their Report, 'Transportation: the Next Twenty-five Years'. This Report has already been discussed as an attempt by the Society to regain what was seen as lost ground politically. Maintenance was, of course, one of the areas over which they felt that they were losing ground. In the Report, the Society outlined their approach to maintenance, centring on two main points. Firstly, they advocated that a clear distinction should be made among the three different types of maintenance: environmental, safety and structural. The need for environmental maintenance was depicted as responding to 'largely perceived needs' which, accordingly, could be carried on reduced to meet financial restraints when they arose. The need for safety was affected by the local conditions and by accident reports which are partly perceived and partly measured. However, according to the Report, the need for structural maintenance was 'real and capable of positive measurement',[58] and was only to be ignored at great peril. So the Report stated the Society's view that while there was room for economies in environmental maintenance, and a little room in safety, there was no scope at all for savings in structural maintenance. The report goes on to acknowledge an important weakness in the surveyors' case: 'The Society recognises that their view on the matter is based on their collective judgment of conditions observed on the ground unsupported by co-ordinated systematic evaluation'.[59]

In an implicit reference to the Road Condition Survey, they announced their willingness to co-operate with the DTp in a study. Nevertheless, they stressed that their professional experience remained an important guide and would 'continue to assert that our considerable store of collective experience provides valuable and reliable guidance'.[60] Secondly, they wanted to see the proper establishment of standards in structural maintenance, standards which were based on measured and proven criteria which could not simply be changed in periods of financial constraint. Unfortunately, the Report did not go on to specify how such standards would be enforced. Presumably, and this must have occurred to them, legislation of certain minimum standards would have been required.

THE RESPONSE TO THE RESULTS OF THE NRCMS

There was a further and related expression of concern about the condition of the roads, in the form of a conference called jointly by the Institution of Highway Engineers and the Institution of Civil Engineers, in December 1978. No doubt it was at least partly inspired by the county surveyors who formed a majority on the IHE Council. Though too there was growing interest among contractors and others in the private sector. Their interest had developed as the roads programme had begun to decline and they increasingly turned to maintenance as offering a large and relatively predictable market for materials and equipment that would remain in existence for the foreseeable future.

The 'Colloquium', as it was rather grandly called, was held in London over a day and a half. One hundred and forty delegates attended, drawn from national and local government engineers and elected members, research organisations, road user organisations and the road construction industry. As is usual with this type of gathering, the main speaker was a government minister, this time John Horam, the Parliamentary Under Secretary of State in the DTp. He made the announcement that there would be a small increase in the money central government made available for maintenance. He went on to stress the value in his eyes of the NRCMS which would improve decision-making by providing 'more objective standards'.[61]

The other speaker from the centre was a civil servant, Trevor Hughes, the Deputy Secretary (Roads and Local Transport), who made a particularly interesting contribution. He explicitly acknowledged the apparent conflict between central and local government but denied that this had a basis in fact. He pointed out that the government had in recent years accepted virtually fully the maintenance bids of local authorities. He also referred to the NRCMS, the results of which were only revealed in part at the Colloquium. As Hughes saw it, the Survey was not as 'blunt an instrument' as had been suggested but was a valuable tool in determining expenditure. He went on to issue a warning to the local authorities:

> Unsubstantiated exaggeration does not help decision-makers, facts do. Both central and local government must give the right degree of weight to the results, both central and local government must use them for the allocation of resources, but local highway authorities must not make use of the results as a case of special pleading if their roads are below the average. If they were below, others were above.[62]

The other main paper at the Colloquium was by Colonel Leech, Chairman of the NRCMS Executive Group, unveiling the results of the NRCMS in public for the first time. Leech was careful to stress that, as yet, the results had to be seen as very provisional as they covered only three years' results. So far the NRCMS had revealed that there had been little change in the condition of all types of road. Rather worryingly for the engineers, there did not appear to be any fall off in the state of the roads as maintenance expenditure had fallen. Colonel Leech argued that this failure could have reflected improvements in efficiency, which enabled less money to be stretched further, and that there would be a time lag between the cuts and their becoming apparent in the Survey. So he doubted whether the government should as yet use the Survey to allocate resources. At the same time he asked the counties and his fellow engineers to be patient; it would be 'unfortunate if there was any attempt to try and discredit the Survey because the answer has not come out as some may have expected.'[63]

The report of the discussion on this paper notes that there was 'clearly some disappointment in the results of the survey'.[64] This disappointment was shared by many present - elected members, officers and representatives from the construction industry. There were some attempts to redefine the process of deterioration in roads. The Kent County Surveyor, for example, argued that the decay of roads should be seen in terms of a curve. In the first three-quarters of the curve there was little apparent deterioration but the last quarter of the curve fell rapidly as the road approached failure. Consequently, at the point at which it was most economic to extend the life of the road, there was little physical evidence of decay. This, what might be called 'catastrophe theory' of road maintenance, became one of the major accounts of why the NRCMS results were potentially misleading.[65]

The controversy over maintenance and the NRCMS deepened in 1980 when the full results of the first four Surveys were released. The results were not encouraging for the professionals. Quite simply:

> After four years of the Survey, it is beginning to be possible to identify some national trends. The picture varies according to road class and there are fluctuations in defects within these classes in a number of cases. Nevertheless, the overall result does not support the claim that the nation's roads are getting worse and, moreover, there are some indications which show a real reduction in some defect levels.[66]

These results were hedged around with qualifications - 'at this stage it is vital not to draw firm conclusions prematurely'.[67] The time period was too short to demonstrate a firm trend; any recent reductions in spending might have led to temporary patching covering up serious longer term defects.

After the publication of these results, Municipal Engineering did a brief survey of the county surveyors' views.[68] The Journal found that there was widespread concern with the way in which the government might use the results in justifying further cuts in maintenance spending. The county surveyors were alarmed too that the results did not concur with their own view of what was happening to the roads. The County Surveyor of Hampshire, and at the time a Vice President of the Institution of Municipal Engineers, was quoted as saying:

> The facts as presented in this report don't bear out what my gut is telling me. I am very concerned that premature decisions could be taken as a result of which it could be very damaging.[69]

The surveyors spoken to by Municipal Engineering concurred with the view that the results were distorted by counties resorting to cheaper, short repairs.

The Survey results, then, seriously worried engineers within both local authorities and the Department of Transport. Despite the fact that, as one county surveyor said, 'all of us were convinced that the Survey would provide a case (for maintenance expenditure)',[70] it had failed to do so. Moreover, since the local government engineers had made the final decision on the Survey design, they could not easily disown its results. The NRCMS results appeared to the surveyors to raise two types of difficulty. Firstly, that of persuading the DTp of the need, as they saw it, to halt the decline of the roads. The difficulty was, they felt, that it would be hard to combat the 'psychological effects in the higher reaches of the civil service'[71] of the NRCMS results. From their contacts within the civil service, they knew that they had good reason to be concerned with the senior civil servants' response to these results. Secondly, the surveyors were worried by how the results might weaken their case in budgetary battles back in their counties. There was more than one story in circulation about county treasurers waving copies of the NRCMS in front of their county surveyors.[72] Meanwhile, the new Conservative

Government, strongly committed to reduce the level of public expenditure, was looking around for services to cut. So, at a time when the engineers faced a serious threat, they found their position seriously weakened.

The fears of the engineers seemed confirmed when the DTp gave its evidence to the Armitage Inquiry on heavy lorries. In its evidence, the Department made it clear that it accepted the Survey results as they stood. It baldly stated in the written evidence that the 'view that the physical condition of roads is declining because of inadequate expenditure on road maintenance is not borne out by any objective evidence known to the Department'.[73] This was a view rejected by the engineers in the Department, as one of them commented, 'there is one faction in this Department which seems to be taking this as good news, who have concluded that this gives them an out'.[74] The Department felt able to ignore the warnings both of its own engineers and of the county surveyors.

Early in 1980, after the release of the 1979 NRCMS, the British Roads Federation approached the County Surveyors Society to ask for their support in arranging a conference on maintenance. The Society's response was cautious. The CSS President's Committee discussed the proposal but decided that it could not be seen publicly to ally itself with a pressure group.[75] Such an alliance might be seen as compromising their professional neutrality. It is worth stressing the point that public sector engineers are careful to maintain some distance between themselves and the private sector; for example, there is considerable reluctance to elect private sector presidents of the IHE among county surveyors. However, the Society finally formed a joint informal working party with the BRF to arrange the conference.[76] So, in the event, the BRF went ahead with the conference with the effective support of the surveyors.

The conference took place in June 1980, its main purpose being to discuss the NRCMS.[77] A much wider range of people attended this conference compared with the earlier IHE/ICE Colloquium. There were over one hundred and sixty delegates, including MPs, elected members, civil servants (including Ron Bridle, the DTp Chief Highway Engineer), fifteen county surveyors, a number of deputy and assistant county surveyors, representatives of the construction industry and road user organisations.[78] The range of people and interests present at the Symposium is a good indication of the pulling power of the BRF and the extent to which it has acquired wide acceptance among policy makers.

Tony de Boer, the BRF Chairman, opened the Symposium by warning that it would be a mistake to see the results of the NRCMS as a justification for the past maintenance cuts or for cuts in the future. He pointed out that over the previous six years total maintenance expenditure had fallen by six percent, which underestimated the real reductions as it reflected past increases in trunk road maintenance and obscured the fall of eleven percent in the county expenditure over the same period.

The Symposium was addressed by the Minister of Transport himself, Norman Fowler. In his speech, Fowler referred to the importance of getting 'value for money' from expenditure on maintenance and, in line with his Government's philosophy on local government, he stressed that the level of expenditure on local roads was entirely a matter for local authorities.[79] The clear message was that any talk of national standards was entirely out of court except on trunk roads. While on the NRCMS he thoroughly supported its aim of reaching objective measures and he insisted that the conclusions were not his but those of local highway engineers who had both designed and carried out the work. So it was clear that, in the Minister's view, the engineers had failed to prove their case.

Not surprisingly, the NRCMS came in for considerable criticism at the Symposium.[80] The county engineers were incredulous at the results and could not see how they bore any relation to the road conditions that they had observed in their localities. Colonel Leech, the chairman of the SCHM Executive Group, was hard pressed to defend the Survey that he and his colleagues had designed. It is clear too, from interviews, that he and his other colleagues from the SCHM were facing similar, forceful criticisms back in the County Surveyors Society. But, from the proceedings of this Symposium, the engineers, at all levels of government, were seeking explanations of what were, for them, anomalous results. Several were being peddled at the time. One was that the Survey, being a visual inspection, did not reveal all the structural failures, many of which had been hidden by the temporary expedient of applying a surface dressing.[81] Another, more theoretical reason mentioned earlier, was that roads declined along a 'doomesday' or catastrophe curve, whereby the roads did not gradually decline but suffered structural collapse suddenly and often without warning. Certainly, the confusion engendered by the Survey results in engineering circles suggests that the engineers' understanding of the nature of the roads was not as complete as they had encouraged others to believe. Moreover, a TRRL engineer confided, during this study, his view that the county surveyors were exaggerating the ineffectiveness of 'temporary' measures - surface dressing and patching - which were quite adequate to hold most existing roads together.[82]

So, the Government took little notice of the highways community's protestations. The Transport Supplementary Grant, announced in December 1980, revealed that in 1980/81 highway maintenance was to be reduced by 7% while capital project expenditure was to increase by 13%.[83] This was in line with the earlier 1980 Roads White Paper.[84] As was noted in the previous chapter, this reflected the Government's emphasis on those lengths of motorway nearing collapse and requiring extensive repairs, as well as new roads to relieve them of traffic. For example, the M40 extension to Birmingham was intended to relieve the M1. The Government saw most of this construction and reconstruction work going out to private consultants. The increase in spending on motorway maintenance can perhaps be taken as the result of pressure brought to bear from within the Department by civil service engineers. They had been sounding off the alarm for some time over the state of these roads.[85] Even after the announcement that further funds would be devoted to the repair of the motorways, the civil service engineers were expressing concern. At a joint ICE/IHE Conference in September 1980, the Head of the TRRL Highway Department predicted that, given the age of the motorway network, at least five percent (ie about a hundred miles) would be under repair annually for the foreseeable future. Rather unusually, too, the assistant chief engineer in charge of trunk road maintenance at the DTp spoke from the floor and said that he doubted that funds would be available to renew and repair at that rate.[86] The crucial problem was that the motorways built in the late fifties and the early sixties, were coming to the end of their original design lives. In fact, many stretches, like parts of the M1, had needed extensive repairs before this. One of the most significant factors in this rapid decay was the increase in heavy lorries. The County Surveyors Society Pavement Design Group's evidence to the Armitage Inquiry had pointed out that on an average UK site, the illegal overloading of lorries accounted for a fifty percent increase in road damage.[87]

It was against this background of further expenditure cuts that the results of the fifth NRCMS were released in February 1981.[88] There was little cheer in it for public service engineers. The findings yet again failed to meet up with the hopes, 'the overall result still does not support the claim that the nation's roads are getting worse and there are indications that some road classes are getting better'.[89] it argued again that short term palliatives were probably still obscuring the real structural condition of the roads.

CONCLUSION

This case study has to be understood against the background of challenges to the engineers' claims to cognitive competence. These challenges were experienced by the engineers at local and national level, while their attempts to counter these challenges were similarly directed at policy makers at both levels of government. Whereas the engineers had, in the past, been able to persuade the policy makers relatively easily to accept their professional judgement, they now felt that it was not so readily accepted. In other words, they felt that their accreditation in the eyes of the policy makers was being lost. Not only that, but there was an associated danger that the engineering community was coming to be seen by others as a sectional group pushing their own interests. So, unless they could restore their earlier, and cherished, image as 'neutral' professionals, their credibility could be further eroded.

The response to these threats by the professional community was to codify what had previously been matters of professional judgement or of applying tacit knowledge. Thus, the introduction of road assessment systems and the National Road Maintenance Condition Survey represented moves by the community towards codification to compensate for the perceived loss of support for their claims to cognitive competence. At that time they quite firmly believed that the Survey would vindicate their professional judgement. The Survey was seen, too, as likely to restore their image as neutral professionals, countering the view they were just another sectional interest. This move towards codification was not, at least primarily, directed towards others within the policy community. Rather it was directed at those outside the policy community, in local and central government, who controlled access to public funds. In this sense, the professional response was also that of the policy community which the profession dominated.

Others at the centre, on the border of the policy community but not as closely identified with it, went along with the idea of maintenance standards. They regarded these developments as offering further opportunities for the monitoring of local policies, enabling them to make value for money judgements relevant to grant and agency payments and to iron out the wide variations in local expenditure that concerned them. In contrast, the engineers adopted the notion of national standards to strengthen their arguments for more resources at both the national and local levels. But the notion of standards of maintenance meant that the engineers had to make explicit their view of what constituted an adequate road surface. This has gone some way towards enabling others to intervene and dispute their views, to make an assessment of the likely effects of cuts in expenditure without reference to them. Similarly, with systems such as CHART, the politicians are enabled to fix levels of maintenance more precisely for different types of road - a decision that was previously left largely to the engineers. However, these moves towards codification have had their advantages for the professionals. Standards and assessment systems mean that the profession/politician boundary can be defined more precisely. Once standards have been 'objectively' fixed, the professionals can openly criticise the politicians for failing to provide the resources necessary to maintain these standards yet retain the protection afforded by the claim to be neutral professionals concerned with the public interest. Similarly, at the local level, assessment systems can be used to limit the involvement of the elected members to general decisions about the priority to be accorded to different types of road, and reduce their capacity to interfere with decisions about particular road maintenance schemes. So, when the elected members ask for particular schemes to be given priority, they can be told that the decisions on individual schemes have already, in effect, been taken.

NOTES

[1] See 'Local Transport Grants', Circular 104/73, 7 August 1973.

[2] G.D.Spearing, 'Highway Maintenance - the National Scene', The Highway Engineer, 25, 3 March 1979, p.38.

[3] F.P.Davies, 'The Government's Role in Maintaining Roads', Paper given at the British Roads Federation Colloquium on Road Maintenance, June 1980.

[4] And also with the Transport and Road Research Laboratory. The County Surveyors Society also have representatives on TRRL steering committees.

[5] While maintenance expenditure in Great Britain, excluding lighting and administration, was over £160 million, of which £18 million was spent on trunk roads. Department of the Environment, Report on the Committee on Highway Maintenance, (Marshall Report), HMSO, 1970, p.12.

[6] The Committee's remit was to 'consider and make recommendations on the planning, execution and financial control of highway maintenance work'. Ibid, p.(xi).

[7] Personal interview with former county surveyor.

[8] Ibid, p.69.

[9] Ibid, p.7.

[10] Ibid, p.6.

[11] Seven of the sixteen chapters in Marshall dealt with authorities' maintenance organisations and manpower requirements.

[12] Marshall Report, p.16.
Hedley Marshall himself, in an interview, said: 'What we did was to show how modern analytical techniques could be focussed on highway maintenance problems, what should be attended to, what was needed and the methods of determining the priorities. That was, in essence, what we did'.

[13] Ibid, p.19.

[14] He is the author of Financial Administration in Local Government, Allen & Unwin, London, 1960. At this time he was Associate Director of the Institute of Local Government Studies, and he had been Treasurer of Coventry CB.

[15] Personal interview.

[16] Marshall Report, p.87.

[17] The Report also set out recommended managemnt principles for local authority maintenance organisations.

[18] Ibid, p.91.

[19] Ibid, p.91.

[20] Personal interview.

[21] G.D.Spearing, 'Highway Maintenance - the National Scene', p.37.

[22] The structural maintenance standards proposed by the Report specified, for instance, the conditions under which surface dressing rather than resurfacing should take place; e.g. a road would be resurfaced when more than 30% of the wheel tracks are cracked.

[23] The new Department of the Environment incorporated the former Ministry of Transport in March 1970. See Paul Draper, Creation of the DoE, HMSO, 1977. The Department of Transport was again separated from the rest of the DoE in 1976.

[24] Personal interview with a county surveyor.

[25] DoE, Roads in England and Wales, 1972-73 HMSO, March 1973.

[26] Ibid.

[27] Personal interview.

[28] From £24 million in 1971-72, to £39 million in 1972-73 and to almost £48 million in 1973-74. DoE, Roads in England, 1973-74, HMSO, July 1974.

[29] The cost of bringing these roads up to Marshall standards was estimated at £60 million over five years. G A Leech, 'The National Road Condition Survey', The Highway Engineer, 26, 3 March 1979, p.2.

[30] Road Research 1970, Annual Report of the Road Research Laboratory. HMSO, 1971, p.111.

[31] Ibid.

[32] Ibid.

[33] Ibid.

[34] Transport and Road Research 1976, Annual Report of the Transport and Road Research Laboratory. HMSO, 1976, p.29.

[35] 'National Road Maintenance Condition Survey 1980', A Report of the Standing Committee on Highway Maintenance Executive Group. DTp, 1980, para.21.

[36] Paragraph based on a personal interview with a member of TRRL staff and TRRL Annual Reports.

[37] The vice chairman of the Group was a deputy county surveyor, as was the convention on working parties of this type, and so it included not just county surveyors but their assistants and deputies as well.

[38] The observations and the quote used here are based on personal interviews with county surveyors and members of their staff. But the observations are over-simplified, in all counties decisions on the allocation of maintenance funds among divisions and districts are influenced by incremental considerations; for example, the need to provide work for direct labour forces which may not be based in that part of the county most in need of repair work.

[39] Transport and Road Research 1974. HMSO, 1975, p.27.

[40] Transport and Road Research 1979. HMSO, 1980, p.21.

[41] Ibid, p.29.

[42] Ibid.

[43] Leech, 'The National Road Condition Survey', p.4.

[44] This announcement was confirmed by the Working Party on Public Expenditure which proposed reducing the level of expenditure on trunk road and motorway maintenance by 75% of that previously planned by 1978-79, Public Expenditure to 1978-79, HMSO (Cmnd 6393), February 1976, pp.62-63. In the same year the Department stopped grass cutting on trunk roads except for necessary safety and weed control reasons, DoE, Roads 1975-76, HMSO, July 1976, para.7.

[45] Unfortunately it has not been possible to see this document. It is referred to in British Roads Federation, Road Maintenance, BRF, London, n.d. (c. 1977).

[46] Leech, 'The National Road Condition Survey', p.4 (Emphasis as in the original). Colonel Leech, the County Surveyor of North Yorkshire, was the chairman of the SCHM Executive Group that administered the survey.

[47] 'Minor Rural Roads', Report of a Working Party set up to study the Problems of Minor Rural Roads, ACC, July 1976.

[48] Ibid, p.13

[49] The county surveyors later advocated a similar response in their evidence to the Armitage Inquiry on Heavy Lorries, and to the ACC when they were considering the issue. Interestingly, the planners advocated restrictions on lorries but lost to the engineers in the ACC. Observations based on personal interviews and ACC minutes.

[50] Colonel Leech has stated that the SCHM took the idea from the ACC Report. He also mentioned that 'an outline of the proposals was discussed at the County Surveyors Society and in

March 1976 members agreed to recommend support for the Survey'. 'The National Road Maintenance Survey', p.4.

[51] All the surveyors interviewed were unanimous in thinking this.

[52] Personal interview with an administrative civil servant in a highway division.

[53] Like SCHM, it was composed of local and central government engineers, a TRRL scientist, and accountants among others.

[54] 'National Road Maintenance Condition Survey 1979'.

[55] Leech, 'The National Road Condition Survey', p.4 .

[56] Ibid.

[57] Personal interviews with SCHM members.

[58] 'Transportation - the Next Twenty Five Years', CSS, 1978, Appendix E, para.4.2.

[59] Ibid, Appendix E, para.4.6.

[60] Ibid, Appendix E, para.7.6.
cf too 'Members of the Society, collectively and individually, have been on record for some time in sounding words of warning against what they are convinced is the dangerous folly of neglecting the proper maintenance of roads', para.7.6

[61] Quoted in The Highway Engineer, 25, 3 March 1979, p.11.

[62] Ibid.

[63] Leech, The Highway Engineer, 25, 3 March 1979, p.11.

[64] The Highway Engineer, 25, 3 March 1979, p.43.

[65] For instance, a DTp headquarters engineer referred to it as a likely explanation, in a personal interview.

[66] 'National Road Maintenance Condition Survey 1979', A Report of the Standing Committee on Highway Maintenance, DTp, April 1980, para.41.

[67] Ibid, para.12.

[68] Municipal Engineering, 6 May 1980.

[69] Ibid.

[70] In a personal interview.

[71] Personal interview with county surveyor, member of SCHM.

[72] A London borough engineer mentioned these stories, with some amusement, to the author. He felt that his colleagues in the counties were finally beginning to experience the rigours of working with more political and critical elected members; something he felt he had been experiencing for years.
It should be added that the results were broken down into counties but each county's results were identified only by a code number in the NRMCS Reports. Individual counties were told their code number. Engineers in the DTp, including the regional controllers, knew the individual county results - a reason for some unease among surveyors.

[73] Quoted in Municipal Engineering, 17 June 1980.

[74] Personal interview with a DTp headquarters engineer.

[75] Personal interviews with county surveyors.

[76] Personal interview with county surveyor.

[77] Tony de Boer, Chairman BRF, in his 'Introductory Remarks' in 'Road Maintenance 1980, Summary and Discussion', BRF Symposium London 2/3 June 1980. In addition to the Minister's speech, eight further papers were given at the Symposium. Three by people who were involved with the NRMCS, Colonel Leech, a DTp statistician and a third tier officer from a county surveyor's department. And the five other papers were contributed by a DTp/DoE Regional Director, an elected member, the head of the TRRL Construction and Maintenance branch and two county engineers.

[78] See 'Road Maintenance 1980 - Delegate's List'. BRF, June 1980

[79] 'Road Maintenance 1980, Summary and Discussion'.

[80] Ibid.

[81] Ibid.

[82] In a personal interview.

[83] Reported in Municipal Engineering, 13th January 1981.

[84] Policy for Roads: England 1980, HMSO, June 1980, para.98.

[85] Personal interview with DTp headquarters engineer.

[86] Reported in Municipal Engineering, 7th October 1980.

[87] Report of the Inquiry into People and the Environment, HMSO, 1980.

[88] 'National Road Maintenance Condition Survey 1980', February 1981.

[89] Ibid, para.27.

8 The Housing Services Advisory Group

INTRODUCTION

This chapter is complementary to the final sections of chapter five, that dealt with the growth of central government interest in housing management during the nineteen-seventies. Among the policy initiatives taken at that time, one of the most important was the creation of a new standing committee concerned with housing services and not just with housing management. In this chapter, the creation of this new Housing Services Advisory Group will be described and, in particular, the backgrounds and allegiances of the participants. These represented a fair cross section of the housing policy community and illustrate the wider range of participants involved than was the case with the roads policy community. As the work of the Group is discussed, the different relationship that held between professional and politician in housing policy, compared with that in roads policy, will become clear. And HSAG is especially interesting as a case study of how professionals, politicians and civil servants have worked together within a policy community. So, by way of a conclusion to the chapter, the problems of working together and the aims of the Group will be reviewed.

THE CREATION OF THE HOUSING SERVICES ADVISORY GROUP

One of the main products of the review of the housing management advisory service within the Department, and of ministerial interest in a new central government housing initiative, was the Housing Services Advisory Group. The delayed response of the Department to Charles Webb's report on the future of such a service was to call a series of two meetings with interested parties to discuss the form that any new central advisory function should take. The first meeting, in May 1975, was with representatives of the local authority associations at elected member level. From this meeting, the Department obtained agreement, in broad terms, to the idea of some form of central advisory service. The next step was to convene a meeting of informed people to discuss the proposal in more detail. This meeting was held at the then recently opened School for Advanced Urban Studies, attached to the University of Bristol.[1] The meeting, held over a weekend, was billed by the DoE as a strictly informal and unofficial gathering. A wide range of people were invited to the meeting, including elected members (interestingly these were selected on their reputations as having something interesting to say, not on the more usual basis of local authority association representation), chief housing officers (selected on a similar basis to the members), civil servants and housing academics.

The meeting did not have a well defined agenda, it appears to have been a general discussion of current housing issues and the possible contribution that a Departmental advisory service could make. It seems reasonable to conclude that the civil servants used the meeting as a sounding-board for views and also as a means to get to know those prominent in the housing field.

After this meeting, a new division was created within the Department with ministerial backing, to specialise in housing management matters.[2] Previously, these matters had been the responsibility of another division, H2, which had dealt with all the 'social' aspects of housing as well as housing management. Once the new division, H11, was formed and the new assistant secretary heading it in post, the people in the Department moved to set up a new advisory group within the Department. No one in the Department favoured the re-establishment of the Central Housing Advisory Committee or a similar type of committee - in their view it had been too large, formal in its operations, and its reports too official. The new committee was to be smaller, informal and less of a separate power base than CHAC had been. This was more appropriate to the approach favoured by the civil servants - they wanted the new advisory group to come up with quick, topical reports, rather than the elaborate type of report that CHAC had produced. Unlike CHAC too, the new Group was part of a specific policy initiative, aimed at improving the standards of local government housing management and administration. But it was also seen as something of a stop-gap before full time advisers were appointed within the Department.

In selecting the members of the new Group, the minister and his civil servants broke with the usual conventions for joint local-central government committees and working parties. Leading elected members from the local authority associations were not automatically co-opted onto the Group. The associations were asked for nominations but it was made clear that these were not in any way to be binding on the Department. The Department officials were anxious to prevent the Group becoming too politicised, and they saw this as inevitable if those appointed were association representatives. As one civil servant observed, 'only a body which is not representative could avoid becoming a political platform'.[3] So several of the Group's members were not leading figures in the associations; for example, Alan Roberts, the then Chairman of Manchester Housing Committee. As one civil servant put it, 'people like Alan were picked up by asking around among those who had their ears to the ground'.[4] Similarly, the housing officers were not all advisers to the associations, though the majority were. Nevertheless, although nominations were not automatically accepted from the associations, some care was taken to ensure that all the associations had some representation on the Group. But the Chairman of the ADC Housing Committee was only added to the Group, after the other members had been appointed, when the ADC protested that it was not properly represented. In fact, from the first, the ADC was unhappy with the idea of such a Group which did not correspond to the usual conventions of central-local relations. Finally, it should be noted that, in the letters of appointment, it was made clear that the appointments to HSAG were in a personal capacity and not as representatives of any association or organisation.

The new Housing Services Advisory Group, as it first met in 1976, was composed of sixteen members. Of this sixteen, four were civil servants - including the Chairman, who was the under secretary in change of the housing directorate which included responsibility for housing management policy. The other three civil servants were the assistant secretary from H11, a regional controller (housing), and Bill Woof, who joined later when he was seconded to the Department as the first Housing Services Adviser.[5] Most of the other HSAG members were from local government. There were three housing officers: Jack Balchin who was then the General Manager of Stevenage Development Corporation but had been Senior Assistant Director of Housing in the GLC; Harry Simpson, another local government housing officer with a strong connection with London who was then the GLC Director of Housing and had been the Director in Lambeth. Balchin had resigned from the GLC, unable to accept Conservative policies,[6] and Harry Simpson enjoyed considerable standing among Labour politicians.[7] The third officer, Malcolm Lee, was then the Housing Director of Nottingham and the President of the Institute of Housing. Presumably his appointment represented a recognition by the Department of the professional association, as well as bringing a provincial viewpoint to the Group. There was one other local government officer on the Group, the Deputy Chief Administrator of Avon County Council, who had some experience of housing and whose main contribution was on legal points.

Another link with both London and the housing profession was formed by Tony Collinson. He knew and admired Harry Simpson, under whom he had served and later succeeded as the Lambeth Housing Director. But, at the time of his appointment to HSAG, he was a Deputy Chief Executive of the Housing Corporation. The other member of HSAG from the housing association sector was Richard Best, the Director of the National Federation of Housing Associations.[8]

The three elected members of the Group did not reflect quite the same emphasis on the metropolis. There were two Labour elected members. One has already been mentioned, Alan Roberts. The other Labour councillor was Donald Hoodless, then Chairman of Islington London Borough Housing Committee and of the London Boroughs Association Housing and Works Committee. Hoodless was a professional housing man, as well as a politician. He was the Director of Circle 33 Housing Association, an association well known for its experiments in co-operative housing. The third councillor, Frank Bushell from the Vale of the White Horse District Council, was the Chairman of the ADC Housing and Environmental Health Committee. As already mentioned, he was appointed as a result of ADC pressure, though he was not a typical backwoodsman Conservative.

Reg Freeson, the Minister of Housing, had been anxious that the tenants should be represented as the consumers of the service. At that time the only association that came close to representing tenants on a national level was the Association of London Borough Housing Estates and so the Department looked towards them for a possible representative.[9] However, the civil servants were rather wary of ALBHE as its leading member was seen by them as a 'committed politician' and not a 'balanced spokesman for the tenants'.[10] But they finally came up with a tenant who was involved with ALBHE but was not seen by them as 'political' in the same way. In the event, he appears to have been generally regarded as ineffectual by the other members of the Group and, in 1977, he resigned from it after buying his own house.

Finally, there was one academic who was recruited onto the Group. Undoubtedly, if Barry Cullingworth had not left the country for Canada, he would have been asked to join HSAG, so considerable was his standing among the participants in the housing policy community. Instead, Valerie Karn, then a lecturer at the Centre for Urban and Regional Studies at Birmingham University was appointed. She had been associated with Cullingworth, having worked with him on an earlier government report on housing and had also been at the Bristol meeting where she had impressed the senior civil servants present.

Before the first meeting of the new Group, the Department prepared a suggested list of topics with which it was thought they should concern themselves. The list was prepared by the civil servants and agreed by Reg Freeson. It was envisaged from the first that the Group should concentrate on producing a series of reports rather than giving day-to-day advice to the Department. The role of HSAG was envisaged as primarily one of giving advice to local authorities not to the DoE. The members of the new Group largely concurred with this view of their role though, as will be seen, there were occasions when they were frustrated in their desire to influence Departmental policy.

THE FIRST PHASE

During its life time, the Housing Services Advisory Group produced eleven reports on various aspects of the housing service, eight of which were specifically concerned with housing management. There were two phases in the work of the Group, each of which lasted about two years. In the first phase, the original Group produced reports in line with a list drawn up by the Department but with the addition of one further report. At the end of this phase, the future of HSAG was reviewed and it was decided to continue but with an enlarged group. This new Group produced another series of reports before it was finally wound up by the new Conservative Government at the end of 1980. In this section and the next, the work of the Group during each of its phases will be discussed in turn.

The first Report from HSAG dealt with 'Housing Training'[11] and has already been discussed, within the context of recent developments in training, in chapter five. However, this Report was never published, mainly because the Department considered that it had been superseded by the work of the City University Housing Training Project. The existence, too, of a published Departmental report was seen as distracting attention from the work of the City University. This highlights an important limitation on the work of HSAG - its lack of any research capability. It could only draw upon what the members themselves had to offer and most of them were highly committed elsewhere with full time jobs.

The next Report was on Tenancy Agreements.[12] At the time this was seen as a 'radical and progressive document' that aimed 'to redress the balance of rights and obligations between tenant and landlord', in the words of one civil servant,[13] though in the view of at least one professional it expressed what was becoming the conventional wisdom. The Report was quite different from the earlier CHAC reports on housing management, it explicitly championed the rights of tenants as consumers and stressed the obligations of local authorities as landlords. Its significance in the debate on the tenants charter will be considered in the next chapter.

As well as using the Group to stimulate policy changes within local authorities, the Department also used it to issue guidance to authorities. HSAG was more likely than the DoE to be able to produce guidance that would be appropriate to local authorities because its members were knowledgeable about authorities and that guidance was more likely to be seen as neutral, locally, because HSAG was only semi-official. So the Department used HSAG as the channel through which to disseminate guidance on Housing Investment Programmes (HIPs). These were introduced during 1977. The new HIPs, which had been considerably influenced by the growing acceptance of the idea of local housing systems, placed considerable new demands on housing authorities in terms of their abilities to gather and process information. They were expected to give a statement of their future housing strategy, a set of statistical estimates of the supply and demand factors influencing new building and refurbishment in both the public and the private sectors locally and a request for capital funding for the next five years. The Department was very aware that many authorities had little notion of how to go about producing this new type of document. So it placed the preparation of advice high on HSAG's list of reports. The Report, the Assessment of Housing Requirements, was published in August 1977.[14]

Early in the following year, the next Report, The Client Role in Public Sector Housebuilding,[15] was published. Again, this reflected a central government concern, but this time over the poor design of local authority housing.[16] The Development Management Group had been set up the year before on a related subject and the two reports were complementary - the Development Management one being a more technical document, dealing with the management and supervision by authorities of building projects.[17] But the HSAG Report focussed on the internal organisation of authorities and, in particular, on the role of chief housing officers. The central recommendation of the Report was that greater efforts should be made by authorities to ascertain the needs and preferences of tenants. The contribution that housing officers could make to this process was stressed - the chief housing officer, according to the Group, should be the major source of advice on social mix, and the provision of amenities, facilities and equipment on estates. Because of this officer's involvement with the day-to-day management problems, it was argued that he was in an ideal position to know about the social implications of different house and estate designs. Moreover, they strongly argued that the whole development process should be under the supervision of one officer and, given the unique knowledge of a chief housing officer, they were the most appropriate officers for the task:

> ...as a major contributor to the brief, the chief housing officer might be made to be responsible for this preparation and issue. The argument for this is that, just as the education officer specifies what is wanted in a school and the director of

social services does the same for a part III home, so the chief officer should be the channel for specifying what is needed in the housing being designed - acting of course under the direction of the committee responsible for housing management.[18]

This bid for increased power and status for chief officers was followed up in the next HSAG Report, Organising a Comprehensive Housing Service later in 1978.[19] This Report, together with the Client Role, represented an ambitious set of claims comparable with those made by planners and social workers several years before. The implication behind them was that things had gone wrong because housing managers had been excluded from housing design which had instead fallen within the jurisdiction of the 'built environment professions'. This, so it went on, had meant that wider, social considerations had too frequently been forgotten in the design of public housing. This view was not restricted to the housing officers. There was a body of opinion in the DoE that felt that at least part of the blame for poor design and difficult-to-let property lay with the design professions.

These two reports dealt with issues over which the Government did not intend legislating. The next Report stepped into an area which had already been the subject of one of the Housing Policy Review Papers - allocation and access to council housing.[20] In fact, the HSAG Report on Allocation and Access could be read as an amplification of that consultation paper, perhaps not surprisingly, since the civil servants serving the Group were anxious that HSAG documents should chime in with Government policy. By this time, both the consultation paper and the HSAG Report reflected a broad consensus within the policy community. It was only in the local authorities and their associations, especially the ADC, that these types of proposal remained controversial.

Both the consultation paper and the HSAG Report were centrally concerned with improving access to public housing for reasons of equity, worker mobility and the most effective use of the housing stock in coping with need. Despite Cullingworth's recommendations almost ten years before,[21] the DoE and HSAG still felt that local authorities were failing to respond to the needs of neglected social groups, such as single parent families who still faced moralistic attitudes from housing authorities. The traditional task of housing provision, building fairly standard three bedroom houses, was no longer appropriate to a changing pattern of need in the community. Another important element in this debate had been the Homeless Persons Act 1977 which, for the first time, placed a statutory duty on local authorities to house the 'unintentionally homeless'. This meant that a wider range of tenant had to be accommodated by authorities. An awareness, too, of the contribution that the allocation policies made to the difficult-to-let problem, led to an emphasis on using allocation policies to respond to the 'need to widen the social make-up of the public sector'.[22] In summary, the changing operating environment of the local authorities was seen, within the policy community, as necessitating a new and more complex reformulation of housing management tasks. Housing authorities had to 'develop policies to weigh different, competing needs'[23] and 'access and allocation policies have to be more sophisticated'.[24] Similarly, on the supply side corresponding changes had to be made to take account of the new categories of applicant, especially by building smaller units and more special housing.[25]

Both documents, then, favoured widening what were seen as restrictive policies on the type of applicant permitted to join the waiting list. But the HSAG Report did not go so far as the consultation paper which argued for open waiting lists. Nevertheless, the Group did criticise, that most persistent of restrictions on housing access, residential requirements, and pressed for measures to increase tenant mobility - a national mobility scheme and appeal machinery to which tenants could turn if they felt that they had been unfairly treated over a transfer request. Their more original contribution to the issue was to put forward several criteria of consumer choice in public housing: limited offers were only justified if what was offered was in line with the expressed wishes of the applicant; that limiting applicants to one offer was never justified; that appeals against allocation should be allowed; that reasons for applicants' refusals should be analysed; and that the same number of offers should be made to all categories of applicant.[26] This was an important recognition of the emerging 'consumer' status of the tenants and applicants, a significant step away from welfare recipient status. The Report also stressed that those applicants for whom authorities had some statutory or other obligation,[27] who had some urgent medical or social reason or who were key workers, should be given priority. Despite HSAG's strong advocacy of improved and more enlightened housing allocation policies, it did not advocate increased central direction but reaffirmed that authorities should be able to form their own policies.

The next Report from HSAG was unusual in that it was initiated by members of the Group themselves, despite the reservations of the civil servants.[28] As one official commented, 'the sub-group thought this up and went away to produce it'.[29] The officials maintained that the subject of the proposed report, single parent families, was not a 'priority issue' and that it would be unwise to produce a report because it might have appeared to pre-empt a circular that the Department was in the process of preparing.[30] This Circular was the government's response to the Finer Report on Single Parent Families.[31] Some of the members of HSAG had already asked that they should be able to see and comment on the Circular while it was in draft form but were refused. This highlights one of the difficulties that the Group faced in attempting to influence policy within the Department - one which the outside members of the Group stressed.[32] The policy divisions in the DoE were not obliged to consult them and only considered doing so over one hundred percent housing matters. Even the officials on HSAG felt that some of their colleagues were unnecessarily reticent about consulting the Group.[33] One aspect of this neglect was the assumption among their colleagues that, despite its carefully chosen name, HSAG was limited to housing management matters.[34] Another was Departmental suspicion of outsiders, especially when some of those outsiders were local politicians.

Several members of HSAG had a strong interest in single parent families. Harry Simpson, that pre-eminent housing officer, had been a member of the Finer Committee and was on record as supporting the more liberal treatment of single parent families by housing departments.[35] Other members of the Group, in particular Valerie Karn, were in touch with the Finer Joint Action Committee, which had been set up by interested people to monitor and chase up the Government's implementation of the Finer recommendations. The Finer Committee had been set up in the last days of the Labour Government but had not reported until 1974. One of the major issues that the Report had considered had been the housing position of single parent families and, in particular, the direct and indirect discriminatory allocation practices of local authorities. The 1977 DoE Circular and the HSAG Report dealt with this issue.[36]

The Report from HSAG considered the problems faced by those lone parents applying for housing and those already housed.[37] It pointed to residential qualifications and points schemes that favoured the standard family as the main obstacles to access for lone parents to public housing. The tendency of authorities to see one parent families as 'inherently less deserving' was condemned.[38] This meant that even when they were re-housed it was often in less popular accommodation - a tendency reinforced by pressures on councillors and officers not to give better accommodation to single parent families.[39] The Report was also concerned with the implications for housing management of the breakup of families and recommended that there should be a more enlightened approach to these breakdowns by authorities as well as improved co-ordination with other agencies. Finally, the Report explicitly related these problems and their resolution to the standing and training of housing officers:

> In so many respects the lack of status of the housing function and hence of housing staff threatens the provision of a good service, by making it difficult to recruit good staff, to release them for further training and, when short-staffed, to spend more time on individual cases.[40]

The Report was not very well received within the Department. The 'civil servants sat on it for ages' and the chairman of the sub-group that produced the Report was 'carpeted' by the civil servants.[41] The latter considered the Report out of line with government policy - it appears that it was feared within the Department that the Report constituted possible ammunition which the local authorities could use in pressing their claims for increased HIPs allocations. However, the non-civil service members of the Group, for the most part, saw themselves as expressing a collective view on government policy and not simply amplifying that policy. The civil servants, in contrast, felt that it was part of their role on the Group to keep it in broad line with government policy, 'if you see inconsistencies, obviously you try to resolve them!'[42]

While this dispute was going on, another sub-group continued with its work on Housing Associations and their Role in Current Housing Strategies.[43] This reflected the growing recognition that housing associations were coming to play an important role in local housing systems - both in terms of numbers and in terms of provision for special groups. The Report, perhaps, should be considered as an attempt to persuade more local authorities to take this view on board and co-operate more closely with housing associations in their localities. The housing associations had been promoted by the previous Conservative Government, which saw them as an alternative form of tenure to public housing for those who could not afford to buy their own homes. The incoming Labour Government continued to support the associations for much the same reasons as the Conservatives. But, at the local level, many Labour councils remained suspicious of associations, as the councils were jealous of their own role as housing providers. The Report was written against this background of local resistance and apparent failure by local authorities to co-operate with associations.

HSAG, echoing the 1977 Housing Green Paper, affirmed the view that 'housing associations have an important role to play in contributing variety and flexibility to the public sector'.[44] But it noted that the attitude of some authorities meant that housing association activity was small or non-existent in some areas. HSAG wanted this uneven spread of activity to be corrected and recommended that funding arrangements be changed to encourage more local authority support.[45] Local authorities could also give more support to housing association programmes for building land acquisition.[46] Finally, local authorities should consult housing associations during the preparation of investment programmes,[47] perhaps setting up liaison groups of elected members, officers and housing association representatives.[48]

After the initial burst of report writing in 1976 and 1977, the pace of the Group slowed down. The impressive output of the Group reflected the enthusiasm of HSAG members who put a lot of their own time into preparing and drafting reports. Most of these early reports were drafted by HSAG members - Tony Collinson and Alan Roberts wrote Housing Training, Valerie Karn Single Parent Families and Jack Balchin Housing Associations. As has already been stressed, the civil servants connected to the Group were also very interested in housing, while the Minister, according to his civil servants and his own account, was strongly in favour of the Group's work.

The members of HSAG had originally been invited by the Minister to serve for two years. So, at the beginning of 1978, the Department decided on a review of the Group. By this time the original list of reports had been completed and there were some changes within the Department that threw into doubt the continuing existence of HSAG. A new Housing Consultative Committee had been formed, though it was mainly concerned with housing finance.[49] In the light of formation of the HCC and the Housing Services Advisory Unit, and the completion of the original remit, the Department was divided over the continuing relevance of HSAG. These doubts were reinforced by the fact that some of those most involved in forming the Group had been transferred out of the Housing Directorate. However, the members of the Group themselves were unanimous in wanting to continue. There was a feeling too among those civil servants involved that to abandon HSAG then would have been seen as indicating a fall off in the Department's commitment to the improvement of housing management and might also weaken the new Housing Services Advisory Unit which was by then establishing itself within the Department.

The review process took some time. There were other pressing issues around, especially the Labour Government's Housing Bill which was being prepared within the Department. There were also criticisms from the local authority associations to be taken into account. These had to be taken on board, especially if the Group wanted to gain more support in order to survive the likely arrival of a new Conservative Government. The ADC had continued to be critical of the lack of association representation on HSAG. At the time of the reappointment of HSAG the ADC 'actually deputed and we had to argue it out'.[50] By this time the ADC criticism appeared more justified as those HSAG members who, when appointed, had held positions in the associations, had subsequently lost office.

THE SECOND PHASE

Under these pressures about another ten people were added to the membership when the new Group was reconvened. Firstly, the local authority association presence on the Group was strengthened by the addition of four new elected members. Councillor John Morgan, who had replaced Frank Bushell as Chairman of the ADC Housing Committee, joined the Group from the ADC; Councillor John Bradley, the new Conservative Chairman of the AMA Housing Committee; Councillor Simon Randell, the leader of Bromley London Borough and the new Conservative Chairman of the London Boroughs Association Housing and Works Committee; Councillor John Mills, the deputy leader of Camden and the Shadow Labour Chairman of the AMA Housing Committee and later, in succession to Bradley, Chairman of that Committee. Unlike the other three, Mills was an established member of the housing policy community. He had been a member of the Housing Policy Review Advisory Group (with Harry Simpson) and the Project Management Development Working Group. Mills was also a member of the Housing Centre Trust Council, as was Harry Simpson and other prominent housing officers.

Otherwise, one new housing director was added to the Group. This was John Park, the Director of Housing in Greenwich and one of the leading advisers to the LBA. Although he was a member of the Institute of Housing (by invitation through the controversial section 16) he had not played a prominent role in the Institute. Interestingly, he had come into housing through working for SHAC[51] and left, after becoming Director of SHAC, to be Deputy Housing Director in Newcastle and had then moved to Greenwich. During his period of working in the North East, he had got to know Bill Woof, the new Director of the Housing Services Advisory Unit.

The opportunity was also taken to co-opt two council tenants onto the Group. By this time the National Tenants Organisation had been formed, in part to press for tenant representation on national bodies like HSAG.[52] The NTO had made direct contact with the Minister, Reg Freeson, and he appears to have asked for their inclusion.[53] The Department approached Irene Bannon, then the NTO Secretary, and asked her if she would submit names. But, in the event, the Department wrote back asking her to join the Group. This invitation may, in part, have been because Irene Bannon was a prominent tenant leader in Reg Freeson's own constituency. The other tenant co-opted was John Gallagher from the Newcastle Federation of Tenant's Associations, which was among the most active in the country, who had been active in the formation of the NTO. There was also one trade unionist, a representative from the GLC NUPE branch, who only attended one or two meetings. Another significant addition to the Group was Nick Raynsford, the Director of SHAC. Undoubtedly this appointment reflects the growing acceptability of housing pressure groups. Though SHAC had also been one of the leaders in the introduction of housing advice centres and this may well partly explain his appointment.[54] Finally, there was also one new HSAG member drawn from within the

Department itself, the head of the Social Research Division in the DoE Housing Development Directorate. This appointment probably reflects the growing importance attached within the Department to the SRD's work on the difficult-to-let problem, a problem area to which the Group was intended to contribute.[55]

One effect of this expansion in numbers was, in the view of many members, to make the meetings more contentious and to reduce the cohesiveness of the Group. Almost all the original members of the Group interviewed felt that the expansion had had a detrimental effect. The job of the civil service chairman became more difficult as he had to control debates among elected members and yet maintain his neutrality. Certainly the changes introduced two new right wing councillors in the form of Morgan and Bradley - both were seen by the rest as 'backwoodsmen'. Interestingly, the other Conservative, Simon Randall, was not identified with these two. He was seen as much more enlightened and several HSAG members commented on how little disagreement there was between him and the Labour councillors.

The new remit of the Housing Services Advisory Group covered a wide range of topics: housing associations, local authority maintenance and repair services, and aspects of the difficult-to-let problem. Again the report writing was done through sub-groups with other people co-opted to give further assistance to these sub-groups. The first report published by the new Group was on Housing Action Areas and the Role of the Housing Associations.[56] This Report focussed on the factors that inhibited the progress of Housing Action Areas and, in particular but not exclusively, the progress of housing associations. Clearly this reflected a departmental concern. Since the Housing Act 1974, housing associations had had the power to participate in the work of housing action areas which were of course declared by the local authority. They had come to make a substantial contribution with over one third of all public sector rehabilitation then being done by the associations.[57] However, work had been progressing more slowly than had been hoped, in some areas work was taking over five years from start to finish. The HSAG Report found that the system of declaring and working on action areas had 'proved basically sound'.[58] So their contribution to assisting with the slippage problem was to list pertinent factors and recommend ways in which local authorities could improve their co-operation with the associations.

The second issue on which the Group reported was the repairs and maintenance service of local authorities. This had been of major concern to tenants associations for years and a National Consumer Council Report had highlighted the issue.[59] The poor repairs service of local authorities was also widely seen, within the Department and outside it, as a contributing factor to the difficult-to-let problem. But by this time, improving this service had become more pressing as legislation was in progress to increase the obligations of authorities to provide an adequate repair service. The HSAG Report recommended that authorities should spell out the obligations of both the authority and the tenant over repairs. It went on to draw attention to the ways in which the tenants could be given more responsibility and influence over repairs. It gave examples of how some, more progressive, authorities were organising their repair systems and getting the tenants more involved and it showed more evidence than the earlier reports, of members of the Group visiting authorities.[60] The Report concluded by reminding local authorities that they should begin to consider giving tenants more responsibility as part of their implementation of the tenants' charter.

Otherwise, the Report considered some of the organisational changes that could improve the local authority repairs service. Not surprisingly, it came down in favour of the housing department being the department to process tenants' repair requests and implementing the maintenance programme.[61] Several members of the Group were drawn to the idea of having workmen based upon estates.[62] This localised service had been pioneered in the housing associations, though a few authorities had experimented with such a service. The major obstacle for authorities was having to overcome union objections for, if such a service were to be viable, the workmen would have to be willing to abandon the traditional skill barriers. The Group was reluctant to venture any discussion or recommendations relating to this issue. One member of HSAG criticised the Labour members for assuming that the interests of direct labour organisations necessarily coincided with those of the tenants but the issue was not discussed in the Report.

There were similarities between this Report and the earlier NCC Report. Certainly, in the absence of any other systematic research in this area, the HSAG Report could hardly fail to be a response to the earlier Report. Both HSAG and the NCC favoured a more decentralised management and repair service, a clearer statement of tenant and landlord responsibility, more tenant involvement on a collective and individual basis, complaints machinery and an integrated housing department. But the Reports differed in their emphases. The NCC Report laid greater emphasis on the contribution that tenant participation could make to the repairs and maintenance services. They suggested ways in which tenants' associations could be strengthened, and repeated an earlier proposal for a levy on tenants to fund local associations and a national tenants organisation. Of course, unlike HSAG, the NCC were aiming their Report towards the tenants as much as towards the housing managers. The HSAG Report was directed towards these managers, both elected and unelected, and incorporates their perspective. So a major theme of the HSAG Report was that of cost - could repair bills be reduced by more planned, preventative maintenance? What organisational changes would reduce costs? The NCC Report was written primarily for the consumers, whereas the HSAG one was written for and by producers. The NCC, accordingly, was worried that authorities might use the tenants charter as an excuse to off-load costs onto tenants. They stressed that large numbers of tenants could not afford to cope with more than some small jobs.[63] Besides this, the Survey in the NCC Report had shown that tenants saw the repairs service as a major advantage of a council tenancy.

The final HSAG Report dealt with Security on Estates.[64] This was directly related to the difficult-to-let problem, a major cause of the difficult-to-let syndrome was widely seen as the high level of fear among tenants for their personal safety and the effects of vandalism. HSAG took the view that the fear of crime among tenants was as serious a problem as actual crime. Some of these crime problems were regarded as the product of the design and layout of buildings, some earlier DoE research had suggested this.[65] Nevertheless, HSAG noted that crime figures were collected at too aggregate a level to enable any firm conclusions on the major factors to be reached. The Report, consequently, did not attempt to diagnose the problem but simply listed proposals that they thought might ameliorate the problem. One proposal was that tenants associations should be encouraged, perhaps through the employment of community workers. Another was that housing management should become more intensive, particular care being taken to raise the level of trust between officer and tenant.[66] Thus training, decentralised management, improved communications between staff, better caretaking services and more liaison between the police and housing officers, were all recommended.

Members of the police had been co-opted onto the sub-group that produced this Report.[67] Some HSAG members felt that the Home Office and the police had been unhelpful in refusing HSAG's request for information relating to the incidence of crime on council estates, not just on the grounds of cost but also on the grounds that the Group had no right to this sort of information. In fact, even some of the information actually given by the police was off the record so it could not be used in the Report.

A further report had been projected by the Group to look into another aspect of the difficult-to-let problem - the 'acceptability of the existing stock'. This sub-group was chaired by Alan Roberts and he had the major responsibility for writing the Report. But, in May 1979, he became an MP. Although he remained on HSAG, his involvement inevitably dropped off, so this Report was never finished.

HSAG was not finally disbanded until July 1980. But from the time the Conservative Government entered office the Group was under sentence. From the general election campaign conducted by the Conservatives it became clear that the new Government would not be sympathetic to public housing. Soon after his appointment, the new Conservative Secretary of State for the Environment, Michael Heseltine, set in motion a wide ranging review of the Department. The review made it clear that the Minister did not think that housing management should be a concern of central government.[68]

The announcement of the Group's abolition was made at a meeting of the Group in July 1980. It was made by the new Minister for Housing and Construction, John Stanley, Reg Freeson's successor. He stressed that the decision not to renew their appointments, which were due to expire at the end of November, was a considered one. It did not reflect on the Group's work or on the calibre of the people:

> The decision reflected the government's view of how relations with local government of housing matters and, by extension, housing management, should be conducted. The main channel of communication on housing management should be the local authority associations, just as the main housing business, such as HIPs, was at present negotiated through HCC and its working groups.[69]

This reflected the Government's view that they should not take a leading role in housing management. Stanley did not reject the idea of some DoE involvement in a housing management issue if requested by the associations. But he saw any future arrangements as involving the use of specialist working parties and not a standing advisory committee, and he instanced the mobility working party then sitting.[70] Moreover, Stanley also saw a Housing Consultative Council extending its discussions to include a wider range of housing issues and acting as the umbrella for working parties. There were further considerations too. Having produced a wide range of reports, HSAG was beginning to run out of issues to report on. Besides, by this time the Housing Services Advisory Unit was an established unit within the Department which made a change of role for the Group difficult.

The announcement had a mixed response. John Morgan, the ADC Housing Chairman was quite happy with the Minister's proposals but he expressed the usual ADC concern that any new working parties should not mean increased costs for the associations.[71] However, Frank Bushell, his predecessor as ADC Housing Chairman, did not agree and felt that HSAG could continue to make a valuable contribution to housing policy. He later wrote to the Minister expressing his misgivings over the abolition.[72] John Mills, by this time once again the AMA Housing Chairman, regretted the loss of a forum where such a wide range of people could get together, but felt that specialist groups could achieve more.[73] Simon Randall, the LBA Housing Chairman, felt much the same and asked for some assurance that the Housing Consultative Council would provide a forum for discussion and that specialist groups would be created when required.[74]

But the housing professionals regretted the passing of the Group. HSAG had proved a valuable asset to the housing profession as a means of gaining some government support for their professional claims and as a symbolic recognition of their growing importance. For the individuals involved, HSAG membership gave them some standing in the professional community and useful contacts in the Department. The professionals, generally speaking, regarded the Housing Consultative Council as a talking shop for the politicians, the highly partisan atmosphere making their contribution all but impossible. The Secretary of the Institute of Housing, himself a former civil servant and HSAG member, protested at the ending of the Group.[75] This response reflected the fear among the professionals that the Housing Services Advisory Unit might be next for abolition. Although, by this stage, it was clear to those involved that the Unit was functioning quite independently of the Group which, despite the earlier assumption, had not become the steering committee for the Unit. According to the civil servants, the main reason for the Unit's survival was Bill Woof's ability to impress the ministers who learnt the value of having expert and sensitive housing advice on tap.[76]

However, the tenants felt that the ending of the Group was a setback for them. Although the reports did not go as far as they would have liked in the direction of more tenant control, it did offer them much needed official recognition and a channel through which to influence housing management policy at the local level.

CONCLUSION

The Housing Services Advisory Group was an unusual standing departmental committee in terms of the range of people it included. One civil servant remarked that it was unique in his experience in bringing together elected members, local authority officers and civil servants in the same committee. The usual convention in arranging central-local government working parties or committees is that politicians deal with politicians and civil servants with officers - this has been well illustrated in the earlier chapters dealing with roads. This convention, then, did not hold in the case of HSAG. Indeed, there was some concern in the Department at the propriety of a civil servant chairing meetings that included elected members and, later on, an MP. It appeared to bring a civil servant too close to matters of political controversy - though, for the most part, the Group steered clear of the very contentious issues.

The civil service members of the Group took a full part in its deliberations. They did not 'necessarily give departmental advice, we would tell them that the Minister was interested in XYZ, but the Group was not doing things under orders'.[77] There was a 'sensible interplay' between the Group and the Department in fixing the former's brief and finalising report drafts.[78] But the civil servants thought that the reports should be broadly in line with government policy, though 'it was pretty rare that we had to scare them off'.[79] Some HSAG members were unhappy about this and felt that they should be able to issue reports that expressed their collective views even when these were critical of government policy. Even so, there appear to have been few instances of conflict,[80] though this may have just reflected the rule of anticipated reactions. But there were also a few instances when the civil servants thought that the Group should deal with some issues and it did not. In the words of one civil servant: 'So sometimes we were pushing people into things. While we didn't see ourselves as policemen, sometimes we had to warn them, but we spent more time urging them into things'.[81]

It was very significant that the Department brought politicians onto the Group as well as housing officers. It was an implicit recognition that housing officers do not enjoy a monopoly of expertise in housing and that many members are as well informed and as knowledgeable as their officers.[82] Whereas too, in many other policy areas, the professional officers and civil servants would have resisted the participation of elected members or attempted to segregate that participation, in the case of housing they recognised that elected member support was essential. Elected members had to be involved if anything was to be achieved in housing policy. In any case, most of the elected members were people whom they already knew and had met in other forums like the Housing Centre Trust, the associations and social events such as parties held by pressure groups. As one chief housing officer noted: 'The people who get involved in these things form a closed shop, they are all people you would know in the normal course of things'.[83]

So most of the HSAG members, especially in the original Group, had met or known each other before. The main exceptions were the Conservative 'backwoodsmen' who, while not unknown to the other councillors, were strangers to the 'closed shop' of housing people who were, at times, more than a little horrified by their views.[84] So the officers, especially the London-based ones, were very used to working closely with the politicians. This was also true of some of the civil servants. Those who were performing or had performed regional office functions had become used to dealing with elected members. The first Chairman of HSAG had had this type of experience, having formerly been in charge of London housing in the DoE. This experience meant that he knew personally about half the HSAG members, both officers and politicians, before the Group was established.

As well as managing the production of reports within the Group, the civil servants on the Group also had to manage the relationship with the rest of the Department. Some of them identified the real problem as being in their relationships with their colleagues in the Department. One of them said that the reports had caused a lot of hostility from the rest of the Department as they were seen by the finance people in DoE as a bid for resources; but that they had had no trouble with the ministers about the reports, in fact, he could not think of any occasion when a Minister had asked for changes.[85]

The other members of the Group had been impressed by the officials' commitment to the Group despite these difficulties within the Department; and they were certainly aware of these conflicts. The other members, for the most part, were impressed by the enthusiasm and commitment of the civil servants. Though they did feel that the latters' knowledge of the grassroots problems of housing management was small.[86] They were conscious, too, of being steered to some extent by the civil servants; as one HSAG member commented: 'In general, the influence of the civil servants on the meetings was moderating rather than political, preventing political issues forming that might threaten the government'.[87]

Insofar as the Group could be said to have had an objective, it was to establish a progressive consensus on local authority housing practices. It avoided issues that were contentious between the two main political parties - such as council house sales, municipalisation and direct labour organisations. Clearly, over issues such as these, the Group itself could not have reached unanimity. But they did not entirely eschew controversy. What they were anxious to reform, the civil servants concluded, were the punitive and restrictive local housing policies which they saw as still present in many areas, especially outside London. So the Group criticised certain existing practices by local authorities, such as the use of residential qualifications, failure to observe statutory obligations, and points systems that discriminated against real need. This criticism angered some local authorities, especially in the non-metropolitan districts. Resentment at, apparently, being preached to by HSAG through their reports, probably helps explain the ADCs hostility towards the Group. HSAG also thought that promoting the professionalisation of housing, through training and enlarging housing departments, would improve these anomalous practices and raise the effectiveness of the housing service locally. This progressive consensus tended towards a technocratic view of the problems and the kinds of solutions envisaged.

NOTES

[1] Founded in 1973 and sponsored by DoE.

[2] Personal interviews. The interviewees were uncertain about precisely when the division was established.

[3] In a personal interview.

[4] In a personal interview.

[5] Sources for membership are HSAG reports and interviews.

[6] 'Balchin, key man under Denington (the former Labour chairman) and a real force for a metropolitan housing strategy, resigned and left County Hall, unable to accept the policies of the Conservative regime'. Ken Young and John Kramer, Strategy and Conflict in Metropolitan Housing, Heinemann, London, 1978. Interestingly, Young and Kramer go on to note that 'senior civil servants' regarded Cutler's policies as 'disastrous for London housing'.

[7] He had left Lambeth to be the first Chief Executive of the Northern Ireland Housing Executive in 1971. According to a senior Labour MP, he was recommended for the job by a Labour Parliamentary spokesman.

[8] The NFHA represented the interests of the housing associations as well as organising training courses and other events.

[9] The growth of the tenants' movement is discussed in chapter 9.

[10] In a personal interview.

[11] HSAG, ' Training for the Housing Service', DoE, August 1977.

[12] HSAG, Tenancy Agreements, DoE, 1977.

[13] In a personal interview.

[14] HSAG, The Assessment of Housing Requirements, DoE, August 1977.

[15] HSAG, The Client Role in Public Sector House Building, DoE, January 1978.

[16] See chapter 5, where the ministerial and civil service concern with this is discussed.

[17] DoE, The Management of Local Authority Housebuilding Programmes, Report of the Development Management Working Group, November 1977.

[18] HSAG, The Client Role, para. 3.10. Interestingly, the increased involvement of the housing officer had already been suggested in an early DoE Report, The Estate Outside the Dwelling, Reactions of Residents to Aspects of Housing Layout, Design Bulletin No 25, HMSO, 1972. This is one of the earlier indications of a growing interest in the DoE in problems on housing estates.

[19] HSAG, Organising a Comprehensive Housing Service, DoE, 1978.

[20] DoE, 'Housing Policy Review: Housing Management: Access and Allocation Consultation Paper', (HPR/MAN[77]) 1 October 1977.

[21] CHAC, Council Housing: Purposes, Procedures and Priorities, Ninth Report of the Housing Management Sub-committee of the CHAC, HMSO, 1969.

[22] DoE, 'Housing Policy Review', (HPR/MAN[77]).

[23] Ibid.

[24] HSAG, Allocation of Council Housing, DoE, September 1978, para. 1.2.

[25] Ibid, para. 1.4.

[26] Ibid.

[27 This rather suggests that the Group was concerned that authorities were not fulfilling their obligations under the Homeless Persons Act 1977 and under Sections 111-113 of the Housing Act 1957.

[28] Personal interviews.

[29] Personal interview.

[30] Personal interviews.

[31] Report of the Committee on One Parent Families, HMSO, July 1974 (Cmnd 5629).

[32] Personal interviews.

[33] Personal interviews.

[34] Personal interviews.

[35] He had given a speech to the National Council of One Parent Families in October 1975, published as One Parent Families - The Role of Housing Authorities, NCOPF, 1977.

[36] DoE Circular 78/77. The HSAG Report described itself as 'an attempt to provide a broader context for the Circular'. HSAG, One Parent Families, DoE, March 1978.

[37] Pointing out the increase in the number of single parent families, about 700,000 in 1977, representing 10% of all families. Ibid, para. 2.1.

[38] Ibid, para. 3.15.

[39] Ibid, para. 3.18.

[40] Ibid, para. 7.4.

[41] Interviews with two non-civil service HSAG members.

[42] Personal interview with civil servant.

[43] HSAG, Housing Associations and their Role in Current Housing Strategies, 1978.

[44] DoE, Housing Policy, A Consultative Document, HMSO, June 1977, para. 9.4.

[45] Ibid, para. 2.8.

[46] Ibid, para. 3.6.

[47] Ibid, para. 3.10.

[48] Ibid, para. 4.3.

[49] As the housing cutbacks increased, in the officials' view, the HCC became a forum for taking up adversarial stances rather than for constructive policy discussions.

[50] Personal interview with civil servant.

[51] The Shelter Housing Aid Centre. See chapter 5 for discussion of SHAC's work.

[52] See chapter 9.

[53] Personal interview with NTO activist.

[54] See chapter 9.

[55] In an interview, a senior officer in the SRD claimed that they had 'discovered' the difficult-to-let problem. They were certainly active in seeking solutions through the Priority Estates Project.

[56] HSAG, Housing Action Areas and the Role of Housing Associations, DoE, October 1980.

[57] Ibid, para. 1.2.

[58] Ibid, para. 2.5.

[59] NCC, Soonest Mended, NCC, November 1979.

[60] HSAG, Organising an Effective Repairs and Maintenance Service, October 1980, para. 7.1.

[61] Ibid, paras. 2.21-2.25.

[62] In a personal interview.

[63] A concern shared by tenants organisations that 'participation' could simply mean authorities reducing their costs by passing on their repair obligations to the tenants.

[64] HSAG, Security on Estates, DoE, November 1980.

[65] DoE, The Estate Outside the Dwelling, 1972.

[66] In a frank statement of the problems of housing management: 'Tenants often find management unresponsive and inaccessible, possibly because of shortage or inadequate training of housing staff, undue pressure and overloading of work and/or poor co-operation between departments. Conflict and mutual distrust between tenants and housing staff may result ...'. HSAG, Security on Estates, DoE, November 1980, para. 4.9.

[67] Ibid, Section 8.

[68] The Ministerial Information System (MINIS) exercise. See chapter 5.

[69] Minutes of the HSAG meeting held on 25th July 1980. HCC is the Housing Consultative Committee.

[70] Ibid.

[71] Ibid.

[72] Mentioned by a civil servant in a personal interview.

[73] HSAG Minutes, 25th July 1980.

[74] Ibid.

[75 He had formerly been the Assistant Secretary heading H11.

[76] Personal interview with civil servant.

[77] Personal interview with civil servant.

[78] Personal interview with civil servant.

[79] Personal interview with civil servant.

[80] Personal interviews with HSAG members.

[81] Personal interview with civil servant.

[82] In fact the civil servants mentioned, during interviews, their high regard for many of the elected members. They were especially impressed by Alan Roberts. Both they and other HSAG members interviewed thought that, in general, it was seldom easy to distinguish between types of contribution made to discussions by professionals and by politicians.

[83] In a personal interview.

[84] This emerged in interviews.

[85 Personal interview with civil servant. He was referring to Reg Freeson as Minister, and not John Stanley, in this quotation.

[86] Personal interviews.

[87] Personal interview with an elected member.

9 The tenant's charter

INTRODUCTION

In chapter five the development of interest in housing management by both politicians and civil servants during the seventies was discussed and related to the support that was given to the profession. While this interest involved, for the most part, indirect attempts to influence housing management practices, this chapter is concerned with another manifestation of that interest which did bring central government to intervene on housing management - the tenants' charter. The origins of the charter, essentially the name given to different proposals to reform the public landlord-tenant relationship, will be traced back to the growth of a tenants' movement supported by other pressure groups. While this movement provided the idea of a charter of rights for council tenants, the political parties will be seen to have adopted the idea and based legislation on it. The measures in the charter meant the central government intervening in matters of housing management that had previously been jealously guarded by local authorities. This chapter will examine the response of the local authority associations and the professional housing assocation to the charter proposals.

THE EVOLUTION OF THE TENANTS MOVEMENT

The first national tenants' association was the National Association of Tenants and Residents, formed in 1948, after the government's abandonment of the wartime rent freeze, to oppose rent increases and to campaign for leasehold reform. So, as its name suggests, it was not limited to public sector tenants but included some owner-occupiers and private tenants. It appears to have been, at least in part, an attempt by the Communist Party to exploit post war housing discontent.[1] NATR continued in existence up until the mid nineteen-seventies but never became an effective national organisation because of its lack of resources and its hard left political line.

In the same year that NATR was founded, the issue of tenant participation came to the surface with the publication of a Political and Economic Planning Tract on the subject.[2] The Tract argued that participation could go some way towards overcoming the bureacratic impersonality of most housing authorities and enable tenants to exercise more control over their own lives. The Tract appears to have aroused no interest in political or official circles as the housing problems at that time were seen almost exclusively in terms of increasing the total housing stock. Against this priority, participation must have seemed very much an optional extra. Even so, the first official recognition of tenant participation was made four years later in a government circular which, in passing, suggested that tenants' associations could help to improve relations between councils and tenants.[3] But further, and more positive, government recognition of tenant participation was not given until a 1959 Central Housing Advisory Committee Report and the accompanying circular.[4]

Down at the grassroots, the formation of tenant associations and federations proceeded very gradually during the fifties and early sixties. The most important federation formed during this period was the Association of London Housing Estates, later the London Tenants Organisation. This was not solely a tenant initiative but was sponsored by the London Council for Social Services, a body financed by the London County Council. In 1957 the LCSS arranged a meeting of representatives from the various estate committees in the LCC area, and about one hundred people from thirty five associations attended. From this meeting the London Standing Conference of Housing Estate Community Groups was formed with a community worker from the LCSS as the first organising secretary. The Standing Conference had, as its main role, the fostering of existing tenants groups and the encouragement of new groups through the provision of advice and guidance. Its explicit philosophy was that groups should form and sustain themselves through self-help. The other major role was as a spokesman for London tenants, both with the local authorities and also as a campaigning body.[5] One of its first tasks had been giving evidence to the CHAC Report, just mentioned, and which recommended increased tenant participation.[6]

The numbers of associations began to grow significantly during the sixties. In London an important stimulus was provided by the reorganisation of the capital's local government in 1963. Thus, apart from the GLC tenants who were dealing with the same type of authority as before, other tenants in London found themselves dealing with larger and more impersonal authorities. There was administrative upheaval, too, as the new authorities attempted to impose consistent housing management policies and rents within their boundaries and tenants felt the changes as their rents increased, and as rent rebate, transfer and exchange schemes changed.[7] Other factors also contributed to the relatively rapid growth of tenant associations in London compared with the rest of the country. Mutual learning and passing on experiences are easier in London where a large number of estates are relatively close together. And the activities of the new Association were strengthened as it had GLC grants enabling it to employ several full time community workers.

Another important factor in the development of a tenants movement was rising rent levels from the late sixties onwards. In 1968 the former Prices and Incomes Board in a report on council house rents recommended that they should be increased and, perhaps rather cynically, it suggested that arrangements should be made for more tenant consultation, presumably on matters other than rents, to soften the blow. The Circular that followed from the Report similarly recommended improved arrangements for consultation and for dealing with tenants' complaints.[8]

But it was the next Conservative Government's proposals for, and then their implementation of, 'fair rents' in the public sector that aroused serious tenant unrest for the first time. The membership of ALHE doubled during the period of controversy following the legislation.[9] Nationally, NATR was revitalised and joined with the ALHE in opposing the 1971 White Paper, Fair Deal for Housing, and then the implementation of the Act. Despite intensive lobbying the 'fair rents' legislation was passed by Parliament, with the Secretary of State for the Environment refusing even to meet with ALHE representatives. Many tenant associations campaigned against the implementation of fair rents but were largely unsuccessful as few Labour councillors wanted to risk being surcharged and the tenants failed to join rent strikes in large enough numbers.[10]

So three main factors can be seen as contributing to the rising tenants' movements - the restrictive and bureaucratic practices of housing authorities, rising rents, and the two local government reorganisations. More generally, changes in other parts of society also contributed. During the sixties and seventies the growth of protest movements in other fields, directed at such issues as planning decisions and homelessness, made protest and opposition to officialdom more acceptable as well as indicating possible protest strategies.

THE ORIGINS OF THE TENANTS' CHARTER

At the national level these tenants' organisations have not come to be a powerful or influential pressure group despite the Labour Party's claims to champion tenant interests. For the most part, the demands of the organised tenants have been expressed in the form of a council tenants' charter. The first mention of this charter came in the form of a private member's bill that was put before the House by Robert Edwards, MP for Bilston, in June 1968. The Bill contained proposals that each local housing authority should be required to set up a housing advisory committee to include tenant representatives which would advise the authority on estate maintenance and the provision of facilities. But the Bill fell.[11] Two years later, Dick Leonard, the MP for Romford, attempted to introduce a similar Bill, the Council Housing (Tenants Representation) Bill. Between April 1971 and March 1973 he attempted to introduce the Bill three times and, on the last occasion, was only narrowly refused a second reading.[12] However, Leonard was assured by Reginald Eyres, the Parliamentary Under Secretary, that the issue would be reviewed by the government once further evidence had been gathered - in particular, by a research study sponsored by the Rowntree Trust.[13] The Department itself commissioned further research on tenant participation soon afterwards.[14]

During this time, the only well funded and established federation of tenant associations, ALHE had begun their own thinking about tenants' rights. In 1970 they produced their own five point tenants' charter.[15] This charter demanded security of tenure on the same basis as that enjoyed by private tenants. At the time it was considerably easier for a public sector landlord to evict tenants, through the anachronistic Small Tenements Recovery Act 1838, than it was for a private landlord.[16] There was a tendency, too, for authorities to issue 'notices to quit' quite soon after a tenant's initial failure to pay. This was intended just as a threat but, ALHE argued, such notices could have the effect of panicking tenants and giving them the impression that they had to move immediately.[17] Clearly, many more tenants were receiving these notices than would be finally evicted. The charter called for the use of such tactics against tenants to be stopped.

Another point in the charter was that public housing should be accepted as a 'non-profit making community service'. In particular, it was proposed that there should be a universal housing allowance which would be financed by the whole community - not by the council tenant sector alone.[18] The charter also demanded representation on housing committees for tenants, though this was seen as only one step towards the objective of estate control by the tenants themselves. Finally, it pressed for tenancy agreements to be more explicit and to indicate the council's responsibilities towards the tenants as well as the tenants' towards the council.

THE FORMATION OF THE NATIONAL TENANTS ORGANISATION

By the early seventies there was still no organisation which could claim to represent the interests of tenants on the national level. NATR had become dormant and ALHE, although flourishing with its own staff, could only claim to represent tenants in London. But at this time two new organisations entered the field. One was Shelter, which had been formed in the mid-sixties as a housing pressure group.[19] The other was the National Consumer Council, which was only formed in 1975 but took up the interests of council tenants as consumers as one of its first concerns. Its reports on aspects of council housing and its financial assistance became the basis for the formation of a national tenants organisation. But before the formation of this new organisation is discussed, the content of these reports will be examined.

Soon after the formation of the NCC, two research projects were begun by the Council's staff - one on rent arrears and the other on tenancy agreements. The report on rent arrears reflected local authority and central government concern with the rising level of arrears and attempted to outline some remedies for authorities which did not involve punitive sanctions, except as a last resort.[20] However, the more influential report was the second one, on Tenancy Agreements.[21] It is worth considering this Report at length, not only because of its importance but also because of the picture it gave of council-tenant relations at the time, as it was based on a survey of tenancy agreements. In fact, a surprisingly high proportion of authorities responded to the NCC's request for copies of their tenancy agreements.[22] Moreover, the Institute of Housing agreed in principle to the research - this suggests that there was some unease in the profession about the agreements that some authorities had with their tenants.

The Report was critical of existing tenancy agreements. The NCC criticised them for being one-sided, unfairly limiting councils' liabilities, being paternalistic, punitive in many cases and incomprehensible to ordinary people. Firstly, the NCC found that councils invariably listed the prohibitions and the obligations on tenants but seldom mentioned the responsibilities that councils had to tenants. Thus, seventy six percent of agreements made no mention of councils' obligations as landlords; for example only eight percent of agreements stated that the council should maintain the structure of the buildings. Meanwhile, the tenants were exhorted to cultivate their gardens (98%), to keep the premises clean, tidy or wholesome (81%), not to be a nuisance to neighbours (94%), not to sublet (43%), not to take lodgers without permission (93%), and not to keep animals other than pets (66%). There were even a small group of councils that attempted to limit their own legal responsibilities through the tenancy agreements.

The second main criticism related to the paternalism underlying agreements, reflected both in their tone and their pettiness. While most of the agreements included sections relating to the keeping of pets and gardens, many councils went beyond general injunctions and included petty restrictions such as no pets in houses and detailed instructions on how to weed gardens. Thirdly, 'a striking feature of the agreements that we looked at is the punitive attitude many councils adopt towards their tenants'.[23] Sixty four percent of the agreements said that the council would terminate the tenancy agreement if it was breached by the tenant. However, the Report derived some consolation from the reflection that if all tenancy agreements were enforced to the letter, then there would be a great many more evictions than there were. In fact, it acknowledged that there were relatively few. 'In general it seems that housing managers and housing committees exercise discretion in the use of their powers'.[24] Finally, the Report was critical of the presentation of the agreements which often used small print clauses and referred to Acts of Parliament that were not further summarised.

The NCC stressed the diversity of agreements across the country and, in particular, to the poor image of the tenant implied by them. 'Council tenants account for 30% of the population and it seems incredible that they need to be bound by such regulations whilst the remaining 70% of the population seem to be able to live peaceably and harmoniously without them'.[25]

The NCC concluded that legislation was necessary to clarify the mutual obligations of tenants and council. This would make it easier for councils to 'prepare comprehensible agreements which were rather more uniform than they are at present'.[26] Any such legislation should also include security of tenure for public sector tenants as there was no reason why private and public sector tenants should be treated differently. In the meantime, the NCC recommended that authorities should begin revising their own agreements. To help with this process, a model tenancy agreement was included as an appendix.

The NCC put a great deal of effort in publicising their Report. In their media presentation they highlighted the pettiness and obvious paternalism of many agreements. They drew attention to the more bizarre inclusions in agreements such as to 'see that the Water Closet is kept clean' or not to 'use the premises for immoral or disorderly purposes'. This type of colourful detail was a godsend to the journalists who gave the Report considerable coverage in the press and on television. The Report and, perhaps even more the accompanying publicity, upset the housing managers. The Institute of Housing reacted angrily, criticising the NCC for naming and apparently picking on certain councils. There was a feeling that the authorities had sent copies of their tenancy agreements along to the NCC in good faith and had then found themselves pilloried in the media.

The NCC Report was followed almost a year later by a more official Report from the Housing Services Advisory Group. While many members of that Group did not think that they were just following in the wake of the NCC,[27] nevertheless, their Report did acknowledge the value of the NCC research, the results of which were made available to HSAG in advance of publication. Significantly, as well, two representatives from the NCC were co-opted by HSAG onto the sub-group that prepared the Report.[28] In this way the HSAG Report could hardly help being a response to the earlier one. HSAG did agree broadly with the NCC criticisms and their recommendations, but added the caveat that 'we are not here concerned with the specific allegations made in the NCC Report about unsuitable or inequitable conditions of tenancy'.[29]

So HSAG wanted to dissociate themselves from the NCC Report. Their preferred approach was that of the insider trying to influence by gentle persuasion. Their criticisms of existing agreements were more muted and there were no references to paternalistic or punitive practices. But they saw many agreements as being unnecessarily restrictive and implied behaviour in which the vast majority of tenants was never involved. They shared the view with the NCC that agreements should be less one-sided and should state the obligations of the council as well as those of the tenant. They followed the NCC, too, in calling for security of tenure for public sector tenants, something that had been in the Labour Government's election manifesto. Related to this the Group felt there should be legal changes to enable councils to amend agreements without having to issue 'notices to quit'. It is worth noting, in view of the Association of District Council's hostile attitude towards HSAG, that its Housing Committee strongly criticised this Report, which had mistaken the 'small print of agreement for the day to day reality'; as far as the ADC was concerned, the 'democratic accountability' of housing authorities was a 'sufficient safeguard against any abuses'.[30]

The interest of the NCC in housing management continued after the publication of the two reports. They initiated further research, this time into the maintenance and repair of public housing, and began to stimulate interest in the creation of a national tenants association. When the NCC staff were working on the agreements report, they 'found that there was no effective consumer movement and we saw it as part of our concern to identify and help to build an effective organisation'.[31] After the publication of <u>Tenancy Agreements</u>, a consultant was appointed on a six month contract to 'identify ways in which <u>Tenancy Agreements</u> should be promoted to tenants organisations and local authorities'.[32] The consultant appointed made contact with the National Association of Tenants and Residents, the London tenants and Shelter. Through these organisations a leaflet, drawing the attention of tenants towards <u>Tenancy Agreements</u> and summarising its conclusions, was widely circulated. The leaflet recommended that tenants' associations should co-operate with each other in persuading their councils to adopt new agreements, clearly stating the obligations of the landlord and giving the tenants greater security of tenure. The consultant suggested that a national meeting of tenants should be organised to promote Tenancy Agreements and the possibility of a national organisation. A national conference was held, sponsored by the NCC and Shelter, with representatives from tenants' associations and it was agreed to form the National Tenants Standing Conference, which was later renamed the National Tenants Organisation.[33]

After the formation of the National Tenants Organisation, the National Consumer Council helped organise a further conference, this time in co-operation with a local authority. This was a one day conference, which was open to anyone interested - civil servants, housing officers, elected members, MPs and pressure groups. It was another follow-up to the Tenancy Agreements Report but this time directed towards the policy makers at both levels of government. The event was unusual in that it focussed on developments in one local authority, Basildon District Council. For the last three years Basildon had been considering introducing new forms of tenants' rights and participation in consultation with their tenants. They began introducing a 'Tenants' Charter' in January 1978.[34] It was not the first local authority to introduce tenant participation; in London, in particular, several authorities had been operating schemes for some years, though very few, if any, devoted the same level of resources and commitment as Basildon. This political commitment seems to have come from Harold Tinworth, the Chairman of the Housing Committee, who took a strong interest in tenant rights and participation.[35] Otherwise, on the officer side, a new principal housing officer, the head of the housing management department, had been appointed with experience of participation in housing redevelopment schemes.[36] Finally, it should be said that Basildon was quite favourably placed, as its housing stock was relatively new and in a good state of repair, and it did not have any difficult-to-let housing.

The Basildon Tenants' Charter was presented to the Conference soon after it had been agreed by the Council. Harold Tinworth, in his introductory remarks, said that the Council was taking the Charter onto the national stage because they wanted to learn from others' experience and to publicise a new approach to tenants' rights.[37] Tinworth explained that one of the main functions of the participation machinery was to involve tenants in the design of new housing, to arrive at a repairs service acceptable to tenants and to keep them informed about the reasons for delays in repair work. Otherwise, the tenants were to be given as much freedom as was possible under the existing law, to improve their dwelling, to redesign bathrooms, to do their own decorating and, under some circumstances, to have materials provided by the Council. A vital part of the Basildon approach was the employment of community development workers to work with the tenants.[38] Not all of these points were new but it was new to include them all as parts of a single package - a tenants' charter.

The NCC also used the occasion to put over their views on how the housing repairs service should be improved. The NCC staff member working on their repairs survey, recommended that councils should provide more information on the time jobs were likely to take and explain delays to their tenants. The Conference also gave a platform to the infant NTO. Dudley Savill, the community work consultant employed by the NCC to assist the emergence of NTO, spoke to the Conference, calling on authorities to set up their own participation schemes and drawing attention to the DoE Consultative Document which called for authorities to establish such schemes.[39]

Later, in the same year, Basildon received further publicity in the form of a BBC Open Door programme, which their tenants had arranged. The tenants and the leading community worker that the Council had appointed fronted the programme. It aroused considerable interest and the Council received a large number of requests from local authorities and other interested people for more information.[40] Both the Conference and the programme made Basildon the tenants' charter showpiece.

THE LABOUR GOVERNMENT'S TENANTS' CHARTER

The Labour Government's Green Paper on housing policy was published in June 1977. The Green Paper endorsed the idea of a tenants' charter as a contribution to improving the quality of housing management, as lifting the stigma from council house tenants, and as overcoming the sense of alienation that contributed to 'problem estates'. It expressed concern at the deficiencies of existing management and the growth of tenant dissatisfaction: 'There is a growing recognition of the importance to tenants of good housing management, but also a widespread feeling that they are not getting it, and that many local authority estates are shabby and poorly maintained'.[41]

The points that the Green Paper envisaged such a charter should include were: the right to security of tenure and to a written tenancy agreement, the right to and to be reimbursed for the cost of improvements when tenancies were given up, the encouragement of sub-letting, and the encouragement of tenant participation and co-operatives. However, the government was not willing to enter into any commitments about more resources being made available. Authorities would have to use their existing skills and resources with increased efficiency, nevertheless, 'consideration is being given to the possibility of a more selective treatment of management and expenditure in the new subsidy system, to take account of the needs of authorities with problem estates'.[42]

These basic proposals were further elaborated in a consultation paper issued in October 1977 to the local authority associations and other interested parties.[43] This paper stressed the importance of 'raising the status' and eliminating 'the second class citizen' position of council house tenants. The Consultation Paper saw legislation as necessary to restore equity between private and public tenants on security of tenure, to even out the differences in practice among authorities,[44] and to ensure that justice was seen to be done over evictions.[45] The NCC and HSAG reports appear to have influenced the Consultation Paper, which advocated tenancy agreements which specified the rights of tenants as well as their obligations, 'based on the guidelines in the Housing Services Advisory Group Report on Tenancy Agreements'.[46] But the Government's provision view was that the content of agreements should not be centrally determined but should be a matter for local decision 'in the light of guidance to be provided by the government'.[47] The Government saw giving tenants more rights to improve their own homes as not just increasing tenants' freedom but also as reducing public expenditure.[48] The Government's view on lodgers was that this should be encouraged in order to reduce under-occupation and to provide more housing for single people. Finally, the Government was worried that high standards of maintenance had become harder to maintain as the 'wider role of the public sector in housing leads to greater diversity in both the physical characteristics of the housing and the people whose housing needs it meets'.[49] Local authorities were exhorted to improve the supply of properly trained and qualified staff, in particular as the 'demand for new development is receding, whilst a greater emphasis is being placed on more refined methods of housing management and a sensitive response to individual needs'.[50] Increased tenant involvement was seen as conducive to improved management but the role of authorities in encouraging participation was regarded as crucial in the light of the 'widespread apathy' amongst tenants. Some possible approaches to participation were indicated but no specific set of arrangements were advocated by the government at this time.[51] The Paper also referred to housing co-operatives - the Minister of Housing's, Reg Freeson, hobby-horse - as a possible way of increasing tenant involvement.

The proposals contained within these two consultation papers were further elaborated in five more papers issued towards the end of 1978. These papers dealt with tenant involvement, tenancy agreements, mobility, lodgers and housing eligibility. The latter four papers, for the most part, dealt with the details of the less controversial aspects of the tenants' charter proposals. The main exception was that on mobility, which suggested that a national mobility scheme should be imposed on authorities - something the latter and, particularly, the ADC had always adamantly opposed.

The Paper on 'Involvement in Management' proved contentious with the local authority associations. It recommended a statutory tenants committee, which represented a definite movement away from the view of the local authority associations and the professional association, towards the view of the Labour Party NEC Housing Sub-Committee and the organised tenants. The Sub-Committee argued that a statutory committee would act as a 'trigger mechanism' in stimulating tenant involvement in authorities that might otherwise do no more than give lip service to tenant participation.[52] This change in approach by the Government seems to have been partly influenced by the NEC's deliberations but also reflected growing ministerial doubts about the real willingness of many authorities to set up voluntary schemes of tenant involvement. Thus the Paper noted that, although some authorities had moved in the right direction, 'many other local authorities on the other hand have shown little inclination to encourage, or even to respond to, the aspirations on the part of tenants to play a constructive part in influencing the decisions which affect their homes'.[53]

The tone of this differed significantly from that of the earlier Paper, which actually mentioned 'tenant apathy' as a major obstacle to tenant participation. The later Paper took a more optimistic line - that the 'involvement of tenants is practical and worthwhile'.[54] The major obstacle now was local authority opposition and the means for overcoming this was the imposition of a statutory tenants committee, 'to ensure that in every local authority there is at least one channel through which tenants can made their views known to the authority at elected member level on matters of importance to them'.[55]

Thus the Paper took the line that the committee was a minimum requirement. It rejected the idea of prescribing arrangements down to estate level, but it did propose that authorities should be under a statutory duty to consider setting up such lower level consultative arrangements. Similarly, the actual arrangements for the statutory committees were to be left up to the authorities themselves, the only requirement being that they should be broadly representative of the different tenants; for example include tenants from the different localities and from different dwelling types. But the issues over which the authority would have been obliged to consult the committee were quite wide-ranging. They included management matters - like changes in tenancy agreements and substantial changes in the administration of the maintenance functions - but also matters such as changes in rent levels, changes in allocation policies and transfer schemes. The tenants committee would have the power to raise any issues falling within the same list, while the authority would be bound to consider any suggestions and to respond to them. However, it was not proposed that any executive functions should be given to the committees.

These further proposals from the Labour Government were widely attacked by the local authority associations and by professional housing officers. The AMA response concluded by describing the proposals as 'ill-conceived' and the Association was quite opposed to a statutory framework which 'will prove a straitjacket which will be counter-productive'.[56] The linchpin of their case against the statutory committee was that the legislation proposed was unworkable rather than that it was an attack on local freedom. They argued that the local authorities had to be left to work out their own schemes in accordance with local conditions and attitudes,[57] on the grounds that unless tenants were willing and ready to get involved in management, elected members prepared to share and consult over their management responsibilities, and officers to accept the right of tenants to have a say and help them practically, effective participation would not happen. In any case, as the AMA saw it, the metropolitan authorities were already in the vanguard of tenant participation. The AMA response referred to the DoE survey, which indicated that more than half of AMA member authorities were already operating some form of participation scheme.[58]

The ADC response was even stronger than that of its metropolitan counterpart. The ADC condemned the proposal as creating just another costly 'level of bureacracy' interposed between the council and the tenants which would impede a 'quick and flexible response to problems'.[59] The Association doubted whether there was great enthusiasm among the tenants for increasing their involvement in management. Those who had 'practical housing experience', 'know' that tenants were uninterested in participation. Given this lack of enthusiasm among the tenantry, the Association argued that there was a serious danger of 'unrepresentative persons or groups taking control of the tenants side'. Moreover, they felt that the duty placed on councils to consult the committee was a challenge to the prerogative of the elected members.

The most developed professional housing officer response came from the Association of London Borough Housing Officers.[60] ALBHO expressed the 'widespread dismay among many housing authorities at the prospect of legislation forcing every authority, large and small, to set up a monster "talking shop"'.[61] In line with the AMA and ADC, ALBHO saw the proposals as imposing a national scheme without considering the diversity of local conditions. The size and nature of the estates within any authority 'meant that there are jealousies and misunderstandings about the benefits tenants on particular estates are thought by others - perhaps in flatted blocks - to enjoy, which will make a community of interest between tenants difficult to attain'. Furthermore, 'tenants, especially those who pay their rents, are notoriously unsympathetic to disadvantaged or "difficult" families and the freemasonry of most tenants associations frequently proclaims itself in petitions to a housing committee, to move a family whose noise disturbs their neighbours'. But the officers were anxious to deny that theirs was a 'cynical response' to tenant involvement: 'everyone in housing management wants to encourage tenants to take an interest, both in the affairs of the estate on which they live and in the upkeep of their own houses ... and the truth - which the Department of the Environment might find hard to accept but which must be stressed in capital letters, is that the majority of tenants do not want to be involved in management'.[62]

THE TENANTS' CHARTER IN THE LABOUR HOUSING BILL

The Labour Government's Housing Bill was finally published late in March 1979, four years after the Housing Policy Review had been initiated. The main part of the Bill consisted of proposals for a new subsidy system for public housing along the lines laid down in the Green Paper. But the rest of the Bill proposed a new framework for public sector tenants' rights. This framework largely followed that proposed in the 1978 consultation papers. Security of tenure was to be given to public sector tenants with the possible grounds for possession now being spelled out in legislation - grounds that included breaking the tenancy agreement and management grounds such as where the council had decided to demolish the building or regain specially designed housing from tenants no longer requiring such housing. Other important rights to be given to the tenants were rights of succession to members of the family who had lived with the tenant, the right to improve and possibly be reimbursed by the council, the right to sublet or to have lodgers and the right to have the necessary information relating to tenancy. In the event of the council unreasonably withholding these rights, the tenant would be enabled to take the council to the county court.

On tenant participation, the proposals contained in the earlier 1978 Consultation Paper were carried over into the legislation. The tenants committee was to be made a statutory obligation on councils. The nature of this requirement was given in some detail - the authority had to ensure that there were at least fifteen people on the committee, that they were broadly representative, that they were not to be treated as a committee of the council, that they were to meet in public at least twice a year and that they were to be provided with accommodation and some officer support. The tenants committee was to be consulted on a wide range of issues, again much the same as those listed in the 1978 Paper. Finally, authorities were to explore ways of setting up more local, estate level, consultative arrangements.

There were two other important, and controversial, measures in the Bill forming part of the tenants' charter. These related to mobility in the public housing sector and had been central concerns since 1945 but, despite this concern, had not become a subject for legislation. The original draft of the Bill proposed the abolition of residential qualifications for council house waiting lists,[63] but this was later dropped by ministers. In addition, the Secretary of State proposed to acquire powers to impose mobility schemes at national and at conurbation level (in the metropolitan counties and London). Under these schemes local authorities would have to set aside quotas of lettings for people needing to move into an area and falling into certain categories like key workers.

THE CONSERVATIVE GOVERNMENT

The Labour Government's Housing Bill fell when the new Conservative Government was elected in May 1979. The election of a Conservative Government was to have a considerable impact on the public housing sector, as the Conservatives set about implementing two major policy commitments - reducing public expenditure and giving council tenants the right to buy their homes at relatively low prices. The so-called 'right to buy' had become 'enshrined as a principal weapon in the campaign to win power'.[64] The Conservatives had recognised in council tenants a group that could be won away from their traditional Labour Party allegiance. They could be appealed to by offering to sell them their homes at cut price. At the same time, the Conservatives could argue that this was a necessary step in reducing a burden on the rest of society in the form of the cost of maintaining a large stock of public housing. Ironically, the stress that had been placed on the problems of council housing, such as problem estates, made it easier for the Conservatives to argue that public housing had been a failure.

The new Conservative Government moved with the speed that came to characterise its whole approach - not least in the Department of the Environment. In May, the new Secretary of State, Michael Heseltine, issued a revised general consent, lifting the restrictions imposed by the Labour Government on sales and allowed councils to give more generous discounts.[65] In July, the right to buy was discussed at the Housing Consultative Council. In the same month the Minister for Housing, John Stanley, gave the first detailed public statement of the Government's legislative intentions in the housing field. Interestingly, he made it during his address to the Shelter National Conference and not at the HCC meeting, to the annoyance of the local authority associations. This was to become another enduring characteristic of this Conservative Government - a habit of making major policy statements without previously consulting the local authority associations. Stanley made it clear that the Government intended to go ahead with legislation on the tenants charter on similar lines to those in the Labour Bill. The important differences were to be the inclusion of the right to buy and the replacement of the statutory committee with a more general statutory obligation for authorities to consult with their tenants. The Conservatives agreed with the associations' arguments relating to differing local circumstances:

... we do not believe that we shall get the most effective and sensible result if the Department prescribes in great detail exactly how that consultation should be achieved. We wish to lay down a statutory obligation on local authorities to consult their tenants, but to leave the authorities free to work out the mechanisms for doing this in order to match the almost infinite variety of local circumstances.[66]

In October, the Department issued consultation papers dealing with aspects of the tenants' charter allowing just four weeks for interested parties to submit their comments.[67] The papers included most of the points in the Labour Bill, plus those mentioned by Stanley. It is worth noting that the Conservatives retained the earlier Bill's requirement that councils would have to levy distress through the county court against tenants, a point which the ADC had opposed at the time of the Labour Bill. But the Conservatives omitted any mention of residential requirements and of any mobility schemes, despite statements referring to the economic importance of the mobility of labour. It seems probable that the Conservatives dropped these measures after pressure from local politicians in their party - during the Labour Government, as has been noted, the ADC opposed the introduction of such measures. There were other Conservative additions to the tenants' charter - the extension of the right to information to cover transfers and exchanges, rent protection for sub-tenants (which was later dropped from the legislation) and an extension of the right to improve to include external as well as internal work.

The response from the local authority associations, all Conservative controlled at this time, was favourable. But the ADC had some reservations. John Morgan, the Chairman of the ADC Housing Committee, commented that the proposals 'would cause significant additional costs and increased manpower to district councils at a time when the government is set on reducing public expenditure'.[68] Moreover, the Association continued with its objection that the tenants' charter simply enshrined what most councils were doing already, it would simply 'result in a field day for lawyers and ombudsmen'.[69]

The AMA and the LBA were both more encouraging. John Bradley, Chairman of AMA Housing Committee, welcomed the 'realistic tone' of the Consultation Paper and felt that account had been taken of the AMA's earlier objections.[70] Simon Randall, Chairman of the LBA Housing Committee, commented that 'any reduction in the role of big brother in the council housing field has got to be an improvement'. The AMA was happy with the general duty to consult, providing that rent levels did not become subject to this requirement.[71] But both the ADC and the AMA disliked the provision for secure tenancies and the changes proposed over the levying of distress. They considered that the new procedures would be cumbersome, time consuming and lead to increased rent arrears.[72] Possession orders were 'a necessary sanction against the tenant who imposes a burden on the community by wilfully refusing to pay his rent'.[73] Similarly, they thought that the new procedure for levying distress would mean losing the immediacy that they felt was important in deterring tenants from building up rent arrears.[74] The two associations also suggested that the Housing Act 1961, which obliged councils to keep dwellings in good repair, should be amended. This change would enable some of the non-structural repairs to be passed on to the tenant, so saving authorities money.[75]

The proposed succession rights were regarded as too wide-ranging by the associations, who felt that succession should be a matter of discretion for local authorities:

> ... the final discretion must be left up to them in order to avoid the situation where newcomers can take over such tenancies at the expense of local applicants. ... The public sector landlord owes a duty not only to the tenant in question but to all his other tenants and to the general public, to see that the best use is made of the housing stock.[76]

Finally, both associations disliked the proposed right of applicants for council housing to inspect the information recorded about their application. The associations argued that this violated the confidential nature of the information that might be given to authorities by social or medical professionals.[77]

Publicly at least, the Institute of Housing was less outspoken about the loss of local autonomy implied by the Government's proposals than were the associations. The Institute welcomed the general principles of the charter. But they added the warning that the implementation of the proposals would lead to 'increases in management expenditure at a time when local authorities in particular are under considerable pressure to reduce costs'.[78] And the absence of a 'detailed statutory framework for consultation' was 'most welcome'. However, the Institute was more positive than the associations in affirming its belief in the principle of secure tenancies. Though they maintained with the associations that the proposed procedure for levying distress would make this approach to arrears ineffectual, so 'there was a strong body of opinion within the Institute which regards this proposal as a retrograde step'. On succession rights, the Institute shared the associations' criticism of the generous government definition of the family. Neither did the Institute favour authorities granting this right when it could lead to serious under-occupation. In this case, authorities should just have to provide suitable alternative accommodation. The Institute made no comment on the right of access to housing files, which had proved a contentious issue within the Institute, though it welcomed the right to information about tenancies as 'constituting good management practice'.[79]

Of course, the most contentious part of the Conservatives' proposed legislation was the right to buy. The proposals in the Right to Buy Consultation Paper reflected the Conservative Government's strong determination to preside over the sale of a large proportion of the public housing stock. Clearly, they had attempted in their proposals to block any possible loophole which might have enabled Labour councils to avoid selling. The idea of a right to buy given to the tenants, which could be enforced against authorities, was essentially intended to prevent Labour blocking tactics.

The Government proposed that council dwellings should be sold, at minimum discounts of one third of their value, to tenants after a qualifying period of three years tenancy, and those tenants with a longer tenancy would be offered discounts of up to a half after a tenancy of twenty years. In order to raise a mortgage, tenants would be permitted to call on up to three members of their household to enter jointly into the mortgage. Moreover, by paying a returnable deposit of one hundred pounds, tenants would be able to keep open an option to buy their homes within the next two years. To prevent too many houses slipping back into the public sector, earlier legislation requiring the inclusion of a pre-emption clause enabling the council to buy back at the original price, was to be repealed. However, if the dwelling was sold within five years of purchase, all or part of the discount would have to be returned to the authority. The procedure whereby the tenant was to obtain the right to buy was to serve a statutory notice on the authority of his wish to buy. The authority was then to be given four weeks within which to respond, indicating whether or not the tenant qualified for the right. In the event of delay or refusal by the local authority, the proposal was for the Secretary of State to be given powers to intervene, to fix the sale. Finally, as an incentive to authorities to co-operate, it was proposed that they should be permitted to add fifty percent of the proceeds from house sale to their housing investment (HIP) allocations, something which the Labour Government had not allowed.

These proposals aroused considerable opposition and controversy. But the three local authority associations all supported the principle of the right to buy, not surprisingly as they were Conservative controlled at this time. But the AMA acknowledged the 'differing views among its members on the government's announced policy, although a majority regarded the right to buy as an overriding principle'.[80] Both the AMA and the ADC were concerned about the categories of housing that were included in the proposed legislation. In particular, they wanted special accommodation for the elderly to be excluded from the right to buy provisions. They foresaw difficulties arising from the sale of flats, which would lead to 'pepperpotting', with owner-occupied flats mixed up with council flats, which would cause management problems and make further redevelopment difficult. Astonishingly, the AMA wanted further exclusion clauses in any Act to protect councils from being sued by the new owners for selling them defective premises. They felt that this could be a serious problem with the 'large number of defects that are known to arise particularly in high-rise and system-built dwellings'.[81] Meanwhile, the ADC was worried at the effects enforced sales could have on rural areas. They argued that sales, especially with no pre-emption rights to buy back, would inevitably mean existing council houses becoming commuter homes, second homes and retirement homes to the detriment of village life.[82] These rural authorities resented being drawn into, what they saw as, a 'big city debate'. This issue of sales in rural areas was the ADC's main objection to the proposed legislation.

The two associations were also concerned with the procedures involved in arranging sales. They were concerned at the proposal that they would be compelled to make loans to tenants, they maintained that they should only be required to make loans on the same basis as building societies. This inability to make 'normal commercial judgements' was made more serious by the proposal that up to three members of the household, and not just family members, would be able to enter jointly into the mortgage. This was seen as risky and as something few building societies would even consider. Significantly, the AMA, but not the ADC, was opposed to the idea of a hundred pound deposit to secure a two year option, thinking that this would lead to abuses.[83] The AMA too had doubts about the proposed sweeping powers enabling the Secretary of State to intervene over any issue relating to sales, which was considered to be a further erosion of local autonomy.

The Institute of Housing evaded the main issue of the right to buy. The Institute did not attempt to criticise the basic principle of the right to buy or defend public housing. In fact, quite the reverse: 'comment is made within the context of the Institute's full support for the principle of owner-occupation and its wish that home-ownership should be available to everyone'.[84] So the Institute went no further than arguing that authorities should be able to fix an 'upper limit on the number of sales in order that they may be able to retain a stock of housing for rent necessary to meet their statutory obligations'.[85] They did not elaborate on this point and the rest of their response was concerned with the conditions surrounding sales, their objections following closely those contained in the ADC and AMA responses.

THE CONSERVATIVE GOVERNMENT'S HOUSING ACT

The tenants' charter proposals in the Conservatives' Housing Bill, apart from the right to buy, were largely bipartisan. The Bill, as it was set before Parliament, made some concessions but not many in the direction of the local authority associations' objections. On consultation, the government further limited the issues on which tenants had to be consulted to just 'matters of housing management', excluding rents. However, the points that were left in should be emphasised. The principle of secure tenancies remained largely untouched though the Government made it easier for councils to evict tenants by removing the provision in the Labour Bill for councils having to go through the county courts, but the Government did not give way over distress warrants. On the right to buy some minor concessions were made relating to the conditions and arrangements of sales, for example, other members of the tenant's family were only enabled to share mortgages if they had been resident for twelve months, instead of the six months proposed in the original Consultation Paper.

In Parliament, the Conservatives' Bill faced stiff opposition from the Labour side on the sales issue. The Labour Opposition were not against sales per se but only against the indiscriminate imposition of sales and the wide powers given to the Secretary of State to enforce sales. Gerald Kaufman, the Shadow Environment spokesman, declared that the Bill constituted a 'fundamental danger to the British political system'. He saw it as giving the Secretary of State power 'to do all such things as appear to him necessary or expedient to enable secure tenants ... to exercise the right to buy and to a mortgage'.

During the passage of the Bill, the local authority associations staged one of their most intensive lobbies of the two Houses of Parliament. The ADC and the AMA arranged numerous briefings for their vice presidents in both Houses, which were given by the advisers and association staff. The AMA in particular, kept up the co-ordinating system that had been established during the earlier consultation stage to respond quickly to amendments and counter-amendments as they arose. This lobbying had little impact. The main results were gaining the exclusion of specialised elderly people's accommodation from the right to buy, and the inclusion of special arrangements for service accommodation attached to social services and educational premises.[86]

CONCLUSION

The idea of a charter of rights for tenants originated within the tenant's movement - though the idea was clearly influenced by the welfare rights approach of the community workers who supported the movement. But, despite their enormous potential constituency, tenant activists and their community work supporters were unable to wield the political influence necessary to achieve their demands. It was only when the Labour Party recognised the electoral potential in these demands, that such support was forthcoming.[87] In part, this was the result of Labour's realisation of the need to have alternative policies to put against the Conservative sales policy.[88] It was also the outcome of pressure from the Labour people in the housing policy community - many of whom were local councillors involved in housing. In particular, this pressure succeeded in committing the Labour Government to proposals imposing a statutory tenants' committee on local authorities. It succeeded, too, in resisting the strong pressure from the local authority associations on the government to drop these proposals. In fact, on both the tenants' charter and the right to buy, the associations only managed to achieve quite minor concessions - apart from the abandonment of the statutory tenants' committee.

The housing professionals were ambivalent about the tenants' charter. On one hand, some aspects of the charter corresponded to what the majority of the professional community regarded as good housing practice. But on the other, they disliked having their freedom of action constrained and were concerned by the problems further participation might create. Their concern with participation related as much to the uncertainties involved as to the threat to their own power. The housing professionals feared that participation would stir up a hornet's nest of complaints from tenants, offer platforms for 'agitators' and, perhaps most importantly, present them with difficult and uncertain problems of organising and managing tenant participation.

The mild and evasive response from the Institute of Housing to the right to buy proposals which intended to reduce drastically the public housing stock, the management of which still remained their main task, calls for some explanation. It can be understood in the light of three considerations. Firstly, the great reluctance of most chief housing officers to risk appearing to take a partisan stance, that is to appear to be aligned with the Labour Party. Secondly, a widespread lack of belief in the positive value of public housing within the professional community, reflecting the failure of that community to evolve a convincing rationale for the existence of public housing. Thirdly, the view prevalent in the professional community was that only a small proportion of the stock would be sold off so it would not pose that serious a threat to the profession.[89] Indeed, many of them took the view that it might even enhance the role of the housing officer as a social welfare worker, for as the better-off families bought their homes, these officers would find themselves dealing with a much higher proportion of 'problem families'.

NOTES

[1] Personal interviews with tenant activists.

[2] Councils and Their Tenants, Political and Economic Planning, 1948.

[3] Ministry of Health Circular 10/52.

[4] CHAC, Councils and their Houses, HMSO, 1959; and Circular 24/59, 1959.

[5] Personal interviews.

[6] Personal interviews.

[7] For the background see Della Adam Nevitt and Gerald Rhodes in G Rhodes (ed), The New Government of London, Allen & Unwin, London, 1972, pp.242-249.

[8] Ministry of Housing and Local Government Circular 37/68, 1968.

[9] Information provided by ALHE.

[10] See L.Skair, 'The Struggle Against the Housing Finance Act', in R.Miliband and J.Smith (ed), Socialist Register, 1975, Merlin Press, London, 1976.

[11] Júlia Craddock, Tenants' Participation in Housing Management, ALHE, 1975, p.13.

[12] Ibid, p.14.

[13] The study reported in Ibid.

[14] Later reported in Anne Richardson, Tenant Participation in Council House Management, DoE Housing Development Directorate, DoE, 1977.

[15] Reproduced in Craddock, Tenant Participation, p.4.

[16] This Act was later repealed in the Housing Finance Act 1972.

[17] Cf. National Consumer Council, Tenancy Agreements, NCC, 1976, pp.7-8.

[18] Quoted from Craddock, Tenant Participation, p.4.

[19] For its critique of and remedies for housing management, see Shelter, Homes for Heroes, 1975.

[20] Behind with the Rent, NCC, 1976.

[21] NCC, Tenancy Agreements, 1976.

[22] In fact 318 out of 365 authorities.

[23] NCC, Tenancy Agreements, 1976, p.17.

[24] Ibid, p.18.

[25] Ibid, p.21.

[26] Ibid, p.21.

[27] Personal interviews.

[28] The two were John Ward, Director of the NCC Social Policy Unit and Chris Holmes, an NCC council member. Two other outsiders were co-opted, Dudley Savill, a community worker, and James Wolley of the Lambeth Law Centre.

[29] Housing Services Advisory Group, Tenancy Agreements, 1977, para.8.

[30] Reported in Local Government Chronicle, 27 May 1977, p.434.

[31] In a personal interview.

[32] In a personal interview.

[33] As reported in interviews with members of NTO.

[34] Basildon DC, 'Minutes of the Housing and Welfare Committee', 25 January 1976.

[35] Interview with the Basildon officer and others. Councillor Tinworth was also a member of the Labour Party NEC Housing Sub-Committee.

[36] The influx of officers into mainstream housing management from working in housing redevelopment areas created a lot of interest in new forms of participation in housing management.

[37] In his 'Introductory Speech', Basildon/NCC Conference; 4/5 March 1978.

[38] Colin Speller, 'The Effect on the Management Structure', Basildon/NCC Conference, Discussion Paper No 4.

[39] Dudley Savill, 'Participation Machinery', Basildon/NCC Conference, Discussion Paper No.3.

[40] Papers provided by Basildon DC on the impact of the Conference and the programme.

[41] Housing Policy - A Consultative Document, HMSO, June 1977 (Cmnd 6851), Ch 11, para.11.

[42] Ibid, Ch 11, para.12.

[43] 'Housing Policy Review. Housing Management. A Tenants' Charter. A Consultative Paper', HPR/MAN(77)2, October 1977.

[44] 'Although most local authorities contemplate eviction only as a last resort, practices can vary between authorities and there have been isolated cases where evictions have occurred over relatively trivial matters, causing hardship and bitterness. Security of tenure would cater for such isolated instances of unjust or over-hasty treatment'. Ibid.

[45] The HPR Paper was especially concerned with the rising level of evictions. 'The total of applications determined by the courts for England and Wales rose from 13,433 in the first six months of 1975, to 22,467 for the same period in 1977. Security (of tenure) would at least clarify the reasons for the increase'. HPR, Annex 1, para.5.4.

[46] Ibid, Annex 2, para.4.

[47] Ibid, Annex 2, para.4.

[48] Ibid, Annex 3, para.3.

[49] Ibid, Annex 5, para.1.

[50] Ibid, Annex 5, para.7.

[51] But it did list authorities, all Labour at this time, that had made progress towards participation. Ibid, Annex 5, para.12.

[52] Labour Party NEC (Housing Sub-Committee), 'The Housing Bill: Proposals and Commentry', November 1978.

[53] DoE, 'Tenants' Rights: Involvement in Management', 1978, para.3.

[54] Ibid, para.5.

[55] Ibid.

[56] AMA, 'Response to "Tenants' Rights: Involvement in Management"', January 1979.

[57] 'Despite assurances to the contrary, it is obvious in reading the paper as a whole that the DoE has every intention of becoming involved in what must essentially be a matter of local arrangement between the local authority and its tenants'. Ibid.

[58] Anne Richardson, Tenant Participation in Council Housing Management, DoE Housing Development Directorate Occasional Paper 2/77, DoE, 1977.

[59] ADC, 'Response to "Tenant's Rights: Involvement in Management"', January 1979.

[60] ALBHO advises the London Boroughs Association.

[61] ALBHO, 'Response to "Tenant's Rights: Involvement in Management"', January 1979.

[62] Ibid.

[63] Residential qualifications had already been removed in London by the London Government Act 1963. But the London Boroughs had reintroduced them through the back door by awarding housing points for years of residence.

[64] A Murie, 'Not So Much A Housing Policy ...?' in P M Jackson (ed), Government Policy Initiatives 1979-80, Royal Institute of Public Administration, London, 1981, p.144

[65] DoE Circular Letter, 18 May 1979.

[66] Speech given at Shelter National Conference, Nottingham, 27 July 1979. Quoted from DoE Press Release.

[67] 'Legislation on House: Tenants' Charter Provisions', DoE, October 1979; 'Housing Bill. The Right to Buy - A Consultation Paper', DoE, October 1979; and 'Housing Bill. Improvement and Repair - A Consultation Paper', DoE, October 1979.

[68] Quoted in Municipal Journal, 19 October 1979.

[69] Douglas Peckham, Director of Housing, Cardiff DC, speaking in his capacity as an ADC adviser to a Housing Centre Trust Forum, reported in Housing Review, May-June 1980, p.102. Several housing directors thought that such a 'field day' would occur, in the event the evidence available suggests that it did not.

[70] Quoted in Local Government Chronicle, 19 October 1979.

[71] Appendix to AMA Policy Committee Minutes, Report No 10/1975, 15 November. This had been in the Labour Bill but was not in the Conservative Bill when published.

[72] Ibid; and ADC, Annex D to Housing and Environmental Health Committee Minutes, 6 December 1979, para.10.

[73] ADC, Annex D, para.10.

[74] AMA, Ibid, para.5; and ADC, Ibid, paras.11-13.

[75] Of course levying distraint is seldom other than punitive, since few tenants in arrears possess enough valuables to significantly cover their arrears.

[76] AMA, Ibid, para.7.

[77] AMA, Ibid, para.23; and ADC Annex D, para.16. The AMA gave the far-fetched example of a terminally ill patient applying for council housing and discovering about his illness in the file.

[78] 'Observations of the Institute of Housing on the Proposed Provisions of the Housing Bill: The Tenants' Charter', 16 November 1979.

[79] Ibid.

[80] AMA, Ibid, para.1.

[81] AMA, Ibid, para.7.

[82] ADC, Ibid, Annex D, paras.8-9.

[83] Frivolous applications and tenants obtaining an option then requesting the maximum number of repairs from the council.

[84] The Institute of Housing, 'Response to "Right to Buy" Proposals', 16 November 1979.

[85] Ibid.

[86] See David McCullock, 'How the AMA Fought - and Improved the Housing Bill', Municpal Review, October 1980, pp.133-4.

[87] Though the idea of a tenants advisory committee had been around in the Party for almost thirty years, having appeared in the 1959 Labour Party Manifesto.

[88] According to some of the civil servants interviewed, it was Peter Shore who wanted the housing policy review to be widened out to include the social as well as the financial aspects of housing. Tony Crosland's original remit for the HPR had limited it to reviewing just the financial aspects of housing.

[89] See, for example, the Presidential Address to the Institute of Housing Conference, July 1980. This view was also put to the author by a number of interviewees.

10 Policy communities and professional communities

INTRODUCTION

A major feature of the development of the British welfare state has been a pronounced tendency towards the professionalisation of service delivery and of policy sectors. While political initiatives have brought new tasks into the sphere of government, the performance and administration of those tasks has been entrusted to members of the relatively new organisational professions. The aim of this study has been to analyse the factors that contribute to, and also inhibit, professionalisation and the emergence of professions as autonomous sources of influence within a policy sector. The starting-point for this analysis has been the two concepts of policy community and professional community and the analysis has proceeded by investigating the relationships between the two types of community in the policy sectors studied.

A 'policy community' is a relatively small group of actors in the policy process which develops to deal with some identifiable class of problems which have or could become the concern of government. These communities include civil servants, senior local government officers, representatives of 'respectable' interest groups and, less usually, local and national politicians. Policy communities are based on shared beliefs and are exclusive, new entrants have to satisfy stringent entry criteria such as having high professional standing or occupying a senior organisational position. Communities also serve different functions at different times in the evolution of an issue. In the case of roads, the policy community became a force for policy maintenance once the government embarked on a major road building programme and a mechanism for closing down access to the policy process. While, in the case of housing, the policy community, during the same period, acted as a force for change and became a mechanism for building up a coalition of support behind that change.

Meanwhile, 'professional community' refers to the leadership group of a profession and not to the whole profession, as in the more conventional sociological usage. The members of a professional community generally know each other personally, are in close touch with each other and possess shared values and beliefs. They are the professionals who are involved in negotiating for support from the representatives of other political and social interests. However, the degree of group solidarity varies among professions. The highway engineers showed a high degree of group solidarity or cohesiveness; the county surveyors, in particular, all knew each other and were able to respond quickly to government policy initiatives because they could anticipate each others' views over a wide range of issues.[1] In contrast, the housing professional community was more fragmented, indeed many prominent chief housing officers did not participate in the activities of the professional association and were divided on many issues quite central to professional practice. Differences in professional cohesion arise from a number of factors. A shared professional culture, based on common training and education, as was the case in engineering, is clearly conducive to cohesion. Whereas the emergence of such cohesion is impeded where key aspects of professional practice are subject to political controversy and involvement by politicians, as was the case in housing.

The professional communities are located within and on the fringes of government. Typically they are associated with policy communities though, as has been seen in the two sectors studied, there are important differences in the relationships between policy and professional communities from sector to sector. The highways professional community effectively straddled and dominated the policy community; entry to the policy community depended largely on the professional standing of the individual within the profession, while the professional engineers were largely accepted as the policy initiators and problem definers. In contrast, the housing professional community, which was far from representing all leading housing officers, existed, for the most part, outside the housing policy community which was dominated by civil servants and, unusually for a policy community, local and national politicians. For housing officers' entry to the policy community depended less on their professional standing and more on contacts with politicians and civil servants.

These differences in the standing and influence of the professions in the appropriate policy communities can be explained in terms of the analyses presented in chapters two and three. Policy communities were conceived as possessing an internal order so that any actors have, in effect, to satisfy certain normative, cognitive and relational requirements in order to gain entry and to maintain membership of a community. Thus, firstly, there are certain prevailing values with which the values of new entrants have to be consonant; secondly, there are assumptions about what constitutes epistemologically acceptable knowledge and, this is where professionalism is particularly significant, a tendency to favour accredited over demonstrated knowledge, rather than undergo the costs of examining the validity of new knowledge. Thirdly, the presence of allies is necessary in order to gain entry to a policy community.

The rest of this concluding chapter will review the research findings. It will sum up the conditions under which professions emerge within government and the changing attitudes towards professionalisation among policy makers; the factors affecting professions' capacity to become an autonomous source of influence in the policy process; the significance of professionalism for the central-local government relationship; and, finally, the role of the professions in policy change and innovation.

PROFESSIONALISATION

The approach to professionalism in this study has followed that of writers such as Freidson and Larson, in stressing the nature of professionalism as a form of occupational control.[2] A 'profession' was defined in terms of an occupation's aspirations towards autonomy at work not just in the sense of the immediate work setting but in the related senses of the ability to be self-governing and to act as an autonomous source of influence in the policy process. The other significant characteristic of a profession, in the assertion of its right to autonomy, is the use of strategies based on claims to exclusive cognitive and normative competences. But it is not enough for a profession to have a strong case simply in terms of its claims to these competences. It must also have support from other significant political and social interests. Thus, an emergent profession has to adapt to the assumptions of these other interests on what constitutes a strong professional claim, in cognitive terms of epistemological acceptability and relevance, as well as in normative terms, of reflecting accpetable and desirable values.

In the governmental context, emerging professions are sponsored by interested and concerned actors. The public service professions in local government have typically developed as a result of sponsorship by groups of reformers and their allies in central government. In other words, the emergence of a profession is preceded by the emergence of a policy community which comes to sponsor the profession. The crucial importance of this sponsoring role of policy communities is well illustrated in the cases of both professions studied. Prior to the development of municipal engineering in the early nineteenth century, two distinct policy communities can be identified despite the rather patchy details available. The first community consisted of social reformers concerned with improving the insanitary and appalling working conditions of the urban working classes. These reformers regarded the professionalisation of the engineering functions, which were added to municipalities' powers under the legislation they sponsored, as essential to the success of their reform programmes. Similarly, the second community was formed by Telford, Macadam and their associates who were concerned with the improvement of road communications. They, too, saw the professionalisation of highway engineering as a crucial part of their reform programme. Both these policy communities shared the view that local vested interests and the limited competence of the local councils of the day were major obstacles to reform. Their conclusion was that the organisation of engineers into a profession would create a countervailing force to these localist pressures. A view that the major public health achievements of the early municipal engineers would seem to have vindicated.

Similarly, but more recently, in the field of housing the policy community took a comparable interest in the development of the housing profession. The sponsorship of the profession followed on from the changing definition of housing problems during the seventies. One aspect of this change in definitions was policy makers' growing concern with the qualitative aspects of housing provision and estate design, and with the management of the housing stock. This change reflected the realisation that serious social problems were arising as a result of poor quality public housing. Another aspect of this change was the redefinition of the national housing problem as a series of local housing problems within both the private and public sectors, requiring policy responses more sensitive to particular local circumstances and needs. Accordingly, local authorities were given more powers of intervention and encouraged to use these powers by central government. These new policy definitions meant that the occupational tasks involved in housing came to be seen as more numerous and complex than before. Within this context, the housing policy community supported, and even created on the profession's behalf, professional claims to be knowledgeable about a wide range of housing activities and, therefore, further professional claims to enhanced organisational power in their authorities. It was the civil servants and the professionals in the policy community who took the stronger interest in the professionalisation of housing. The politicians were more sceptical and tended to see the problems more in terms of resource shortages than did the former.

This phenomenon of policy community sponsorship of an emerging profession can be pointed to in the case of other professions. Planning developed as a profession with the support of those concerned with providing an improved urban or suburban environment, particularly with controlling the spread of urban sprawl.[3] While social work, although its origins can be traced back to the nineteenth century and the Children Act, 1948, received an important boost to its standing and influence in local government as a result of the Seebohm Report recommendations and the subsequent Social Services Act, 1972, making the appointment of directors of social service obligatory on authorities. The Report and subsequent events largely arose from the activities of powerful allies of the profession.[4]

However, the internal dynamic of a profession is also important, as well as outside sponsorship. The aspiring profession itself has to be able to exploit the opportunities opened up by its sponsors. This highlights what was a strength of engineering and a serious weakness of housing. The highway engineers were able to exploit these opportunities and, as they came to dominate the policy community itself, create their own further opportunities for advancement and policy influence. They were successful in formulating an account of the policy problems, the required policy responses and so gaining widespread acceptance of their professional account which, of course, stressed the unique relevance of the profession in dealing with these problems.

In contrast, the housing professionals failed to develop their own accounts of professional practice. Insofar as they can be said to have possessed a professional account, it was contained within the redefinition of housing policy already mentioned and was provided <u>for</u> them and not <u>by</u> them. The new professional account, summed up under the label 'comprehensive housing service', was adopted by the professional community which recognised the leverage it offered in terms of the potential for professional expansionism. The account both reflected the influence of the new managerialism as well as the redefinition of housing policy and gained plausibility because of its association with the new managerialism. Thus, in the aftermath of local government reorganisation, the profession was well placed to argue for the establishment of large housing departments headed by a chief officer. However, the evidence indicates that the comprehensive housing service idea and the enlarged housing departments have been a mixed blessing for the housing profession. While many new prestigious posts of director of housing have been created, many of them have gone to officers from outside the professional ranks.

The real problem for the profession was that the comprehensive housing service ideal represented a quantum jump in terms of the profession's claims to competence. When the concept of the comprehensive housing service became an idea in good currency, housing officers were still seen by others and indeed themselves, for the most part, as essentially managers of the public housing stock. They were not thought to possess the cognitive competences to undertake the new tasks involved. This perceived gap between the competences required for running the new departments and actual competences of housing professionals was widened by the arrival of the new housing investment programme system. The importance of HIPs as major bids for resources increased the reluctance of the elected members and other rival professions to entrust the responsibility for HIPs to officers whose claims to relevant and accredited knowledge were so limited. Unfortunately for housing there were other professions, particularly planning and accountancy, whose competences were seen as more relevant and better accredited than those of the housing professionals. The introduction of HIPs, then, further contributed to the opening up of senior housing jobs to officers from other professions.

But this problem was only in part brought about by housing's weak claims to possess many of the relevant and specific competences seen as necessary. It also reflected the difficulty that all public service professions tend to face as they develop a technobureaucratic branch. The greater the proportion of work time that is spent at a senior level on managerial and co-ordinating tasks rather than distinctly professional tasks, the weaker will be their claim to a monopoly of the new technobureaucratic posts. This situation produces a tendency for the members of the technobureaucratic branch of a profession to adopt their own professional accounts to justify their claims to a monopoly of the senior management positions in their service organisations. Thus the more technobureaucratic a profession becomes, the more difficult it is for that profession to resist the encroachments of other technobureaucratic professions. For the opposite reasons, as long as a profession remains a predominantly public service one, the easier it is for it to resist the encroachments of other public service professions.

This problem was also faced by municipal engineering. After the 1972 local government reorganisation, municipal engineering began to decline as a distinct professional grouping. At district council level, reorganisation removed two major engineering functions, as water became the responsibility of the new water authorities and highways the responsibility of the counties. Although, at first, agency arrangements were fairly common, both water authorities and county councils have withdrawn them in many cases, so leaving the district engineers with fewer and fewer distinctively engineering functions, the main residual engineering functions being local sewers, land reclamation and some land drainage. This loss of functions and, therefore, the weakening of the rationale for the engineering profession at district level explains why the highway agency issue was so strongly contested by the district engineers. So that, in relation to the engineering functions, the importance of other functions and professions like environmental health and development control has grown. The response of the Institution of Municipal Engineers was to propagate a new technobureaucratic account in the form of directorates of technical services. These directorates were enlarged engineering departments which included not only the engineering function but also other 'technical' functions; in particular, planning, environmental health and, sometimes, housing. After reorganisation the great majority of these new departments were headed by officers with engineering qualifications so that initially, at least, the municipal engineers weathered the problems arising for them out of reorganisation. But the broad nature of these directorates has also reduced the apparent need for them to be headed by people with a specific engineering qualification. The growing strength and importance of other rival professions, particularly environmental health in non-metropolitan districts, has meant that engineers are by no means assured of a continuing near monopoly over these posts.

RECENT CHALLENGES TO PROFESSIONALISATION

Since the late nineteen-seventies, the professions in local government have been experiencing a major change in attitude from policy makers in central and local government. Policy makers have shown increasing scepticism and resistance to professional claims which has meant that new claims have been subject to much closer scrutiny and relatively few have been accepted. Until the mid-seventies, local and central government were in an expansionary and optimistic phase and so were receptive to new ideas and willing to take risks and, most importantly, had the resources to do so. Certain professions, like planning and social work, had the tremendous advantage of asserting their claims in this climate favourable to professional aspirations. However, by the late nineteen-seventies, the climate had changed decisively. Public spending cuts and the consequent cautious approach to change as well as some disillusionment with the claims that had been made by the professions, conspired to produce a climate hostile towards the attainment of professional aspirations.

It was within this climate that the housing profession and its allies were advancing proposals for improved professional training and qualifications. In particular, the leading professionals and the civil servants immediately involved wanted to see the establishment of a central housing training council to place the profession more on a par with the main local government professions. But there was little support forthcoming from the Department of the Environment as a whole, because of the departmental concern with the financial implications and, similarly, the local authority associations were unwilling to commit themselves to provide the substantial funds required. Moreover, once the Conservative Government came to power, even lower priority was given to professional claims as the Government took the view that the training of local government officers was a matter for local government itself and not central government. This represented an important break with previous post-war administrations, all of which had regarded the training of local government officers as, at least, partly a central government concern.

THE ENGINEERING PROFESSION AND POLICY

Professionalisation also involves the acquisition by a profession of policy influence. Professional communities become autonomous sources of influence within a policy sector in the sense of being relatively independent of other actors in local and central government. The highway engineers were able to become such an autonomous source, whereas the housing profession remained dependent on other policy actors, particularly local politicians and civil servants. In highway engineering, the professional community made a major contribution towards creating the modern policy community in the roads sector. The engineers' claims to cognitive competence were widely accepted as being well accredited. Thus the professional community provided the problem definitions and policy responses as well as most of the leading participants in the policy community. In contrast, the housing professional community had very limited influence within the housing policy community. That community was dominated not by professional accounts of policy but by politicians' accounts and, to a lesser extent, civil servants' concerns with the orderly local administration of public housing. The professional members of the policy community were in a minority and owed their presence in the community less to the accredited status of the profession and more to personal contacts (that is, they had to have demonstrated their claims to knowledge) or to straight nomination by the local authority associations.

In the case of roads, it is clear that a well organised and co-ordinated professional community can exercise considerable influence over policy. The engineers were successful in obtaining widespread support for their professional account with its emphasis on the key function or roads in transport. As already mentioned, this was possible because the profession was well accredited, the policy community was able to act cohesively and there were no competing policy definitions early on in its development. It also had the powerful roads lobby applying pressure in Parliament and on ministers, as a significant ally in maintaining the political commitment to roads.[5]

This cohesion was strengthened by individual career movements from local to central government as the latter's interest in roads grew. So that, by the 1940s, there was an influential professional community with important allies in government and pressing for the government to adopt a major programme of road construction. The County Surveyors Society, too, was becoming a major influence within the closely overlapping professional and policy communities, in the formulation and implementation of the roads programme. While the creation of the road construction units in the late nineteen-sixties further strengthened professional links between the two levels of government by creating a horizontal organisation. What emerged, then, in the late nineteen-sixties, was a 'self-contained, integrated hierarchy for road programme administration and technical work'.[6]

This self-contained hierarchy had a strong sense of mission and responded to the threat from the growing anti-roads issue network after the mid-sixties. The vociferous environmental and amenity groups comprising this network, had succeeded in delaying the construction of major road projects by several years. So much so that the degree of slippage had made the roads budget very vulnerable to cutbacks. Moreover, the increasing controversy surrounding road building, which achieved dramatic expression in the disruption of public inquiries by protesters, caused the politicians to intervene to compel the engineers to allow more, at least symbolic, participation and to give greater weight to environmental considerations. Both the politicians and the administrative civil servants had become very critical of the engineers' lack of political judgement and sensitivity in coping with opposition to road schemes. As a result, the latter were brought under tighter control, non-engineers were placed in senior posts previously occupied by engineers and the road construction units placed under the purview of the regional DoE directors. The engineers reacted by trying to pressure the Department into issuing a ministerial statement reaffirming its commitment to a roads response to transport problems but unsuccessfully as, by then, this could have proved politically risky for the Minister. At the same time, the engineers in DTp and the RCUs drew up plans to out-manoeuvre what they saw as well organised and concerted campaigns to subvert the road programme. Undoubtedly, in seeing the opposition in this way, they overestimated it degree of organisation and political ambition and underestimated the extent of popular opposition to the roads programme.

Financial restraint and the growing scepticism of professionalism already mentioned as factors impeding the professionalisation of housing, also affected the policy influence of engineering. Financial restraint reduced the roads programme, leading to the abolition of the road construction units, and cutbacks in expenditure on road improvements and maintenance. These financial pressures took effect at both the national and the local levels. At both these levels the erosion of deference towards professionals left the engineers less able to oppose cutbacks. This loss of deference became evident in engineers' relations with the public and their own elected members. As a result, by the mid-seventies, the engineers felt that the 'initiative was going away from us'.[7] They began to change their strategy to take account of the new thinking on transportation, while at the same time maintaining their emphasis on roads. The County Surveyors Society responded to this change in two successive reports on transport policy, intended to defend their continued role in transport policy.

Meanwhile, in the increasingly important field of highway maintenance, the response to the felt loss of deference among the engineers was to codify what had previously been the application of tacit knowledge. This was a response to pressures at the local level and, particularly, at the national level, to reduce maintenance spending. Indeed, as was noted in chapter three, the codification or systematisation of a profession's knowledge base is a typical response to challenges from outside a profession. As such, it reflects the sense within a profession that their members' opinions are no longer accepted as uncritically, on the basis of professional accreditation, as they have been and that lay people are increasingly expecting that those opinions should be demonstrated and not taken on trust.

The introduction of road condition assessment systems was also seen by many county engineers as a means of re-establishing clear political-professional boundaries against the growing 'interference' of elected members in their road maintenance programmes. Codification can also have normative implications; in this case it offered the engineers a way of trying to restore their cherished image as 'neutral professionals'. There was widespread concern at the time, in the professional community, that they were coming to be seen as just a general sectional interest, acting from self-interested motives. The hope of the engineers was that by providing 'objective' evidence to back up their claims for increased spending on maintenance, they would go some way to restoring their image as a relatively disinterested group, concerned with the public interest.

Thus the road condition assessment systems and the National Road Maintenance Condition Survey (NRMCS) were codifications intended to compensate for this felt loss of support for their professional claims to competence. It is clear that both these initiatives came from within the professional community and then gained support from those in central and local government. The former saw codification as consistent with their interest in creating opportunities for increased central financial control and in raising the standards of the administration of maintenance. While the local authority associations went along with the codifications under the influence of their professional advisers.

The irony, for the county surveyors, was that the results of the NRMCS did not support their own dire warnings of the conditions of the nation's roads. These results were contrary to their own expectations as they had, for the most part, been confident that the Survey would strengthen their case. In the event, the opposite happened. The NRMCS was used by others to cast doubt upon the engineers' arguments both locally and nationally; ministers and administrative civil servants, in particular, used it as a convenient excuse for ignoring the surveyors' and local authority associations' warnings about the state of the roads. This situation highlights the risks for a profession inherent in codification. It does mean that the members of a profession, in substituting determinate for tacit or indeterminate knowledge, risk losing control to others over an area of professional judgement.

THE HOUSING PROFESSION AND POLICY

In contrast to the professional engineers, the housing professionals never attained the position of being an autonomous source of influence on policy. Their standing in relation to the housing policy community was only marginal. In fact, the housing policy community was an unusual policy community in that the professionals played a minor role, while local elected politicians and civil servants were the more significant actors. Certainly no local politicians could meaningfully be said to belong to the roads policy community. This unorthodox community membership was well illustrated in the case of the Housing Services Advisory Group, which included most of the leading community participants. In the experience of the civil servants involved it was unique, though those with experience of working in regional offices had already had extensive contact with councillors on a more local level. For the most part the civil servants, politicians and others, worked together remarkably well, illustrating how agreement on intermediate values can sustain a policy community despite differences in basic values. In this case, the intermediate value consensus revolved around the progressive concern that local housing policies were both poorly thought out and administered and that central government initiatives were required to remedy this situation.

The events leading to the tenants' charter legislation, described in chapter nine, point up the very limited influence possessed by the housing professional community. The charter, at least in original intent, promised to be a major reform of the municipal landlord-tenant relationship, yet the professionals themselves had little influence in the debate. Initially, the idea of the charter arose from the welfare rights and tenants movements and the debate took place within those networks and not within the professional community. The latter remained mostly silent on the issue until legislation was actually mooted. Again the charter did not become an object of legislation until the Labour Party recognised the electoral potential of the charter as a counter to the Conservatives' council house sales policy.

The professional housing community was very divided over the tenants' charter. Although some of the leading housing officers shared the concerns of the policy community in supporting the idea of increased tenant participation and less paternalistic management, the majority of them were suspicious and tended to reflect the concerns of their employing authorities. Significantly, the leading professionals reflected their authorities' concerns over retaining local discretion and, consequently, resisted the idea of making good housing management practices mandatory on local authorities. Similarly, when the new Conservative Government tacked the 'right to buy' measure onto the tenants' charter, the professional association, the Institute of Housing, overlooked the threat to the public housing stock and limited itself to a few technical observations.

The housing profession, then, in strong contrast to the engineering profession, never became an autonomous source of policy influence. This contrast reflected major differences between the professions in their standing and influence within the appropriate policy communities. The engineers were considerably more successful in establishing plausible and convincing claims to cognitive competence and in acting collectively to manage relations with those outside the profession.

Cognitively, the great strength of engineering was that its members' claims to competence could be presented as 'technical' or 'scientific' and were generally accepted by others as so. Just as importantly, these claims were well accredited because of the existence of well sanctioned and established professional associations as well as the close connection between the profession and the universities. However, the cognitive claims of the housing profession were more questionable. They could not easily be presented, if at all, as 'scientific' and these claims were generally not seen as accredited. Typically, housing officers had no alternative but to accept that their work was essentially managerial or administrative, particularly as many others such as local elected politicians, saw their work as simply the application of 'common sense' and not of esoteric knowledge.

Turning to the management of external relations by a professional community, the latter's ability to do this is crucially affected by the degree of solidarity or cohesiveness of the professional community. Professional solidarity has been shown to have been an important asset to the engineers, despite the divisions that arose over the highway agencies. This group solidarity was based on the common occupational culture that develops from extensive shared experiences of training and professional socialisation. In engineering, the training is largely entrusted to outside agencies and not to local authorities and this has encouraged a sense of professional identification as opposed to local identification. Whereas, in housing, given the absence until recently of full time higher education, courses in housing, locally based sources of socialisation have dominated in the absence of more cosmopolitan sources. The dominance of these locally based sources in the socialisation of housing officers has contributed to preventing the emergence of anything resembling a professional culture within housing.

There have also been strong contrasts in the contexts within which the two professions have been advancing their claims. The engineers enjoyed a long period of time during which road construction and other roads expenditure gradually increased and remained uncontentious. At first, the type of policy response to the internal combustion engine was a politically contentious matter between those who believed that the proper response was investment in roads and the anti-car lobby which opposed such investment. Once that debate was settled, in favour of road investment, roads became an uncontentious issue. The fluctuations in expenditure that did occur, reflected changing economic factors and wartime conditions, rather than changes specific to the roads policy sector. Certainly, once there was a political commitment to roads, the task of the engineers was the much easier one of defending an established policy line rather than the more difficult one of attempting to establish a policy change.

In contrast, the nature and extent of the provision of public housing has remained politically controversial from the first. There have been fluctuations in the extent of controversy but it has persisted. This controversy has meant that, in housing, unlike many other policy sectors, the professionals have failed to establish boundaries between professional and political jurisdictions. The high degree of interest among politicians in housing reflects partly their perception of it as a field which is dominated by redistributive policies and partly the tendency of local politicians to become involved in the day to day management of the housing stock. These political pressures on housing officers have severely limited their ability to build up a distinct professional jurisdiction. In other words, these political pressures have hindered the formation of a professional account around which the profession could be organised. In the absence of such a professional account, the conception of 'good housing practice' has either been borrowed from the policy community or, more often, housing officers have simply articulated the localist or parochial concerns of their employing organisations.

PROFESSIONS AND CENTRAL GOVERNMENT : ADVICE, REGULATION AND PROMOTION

The division of functions between central and local government has been a major factor in the professionalisation of local government. Professionalisation is typically sponsored by actors associated with central government as a countervailing force to the forces of localism. But, despite their origins, there is no necessary or continuing reason why professions should be instruments of central government policy. They can come to act as autonomous sources of policy influence, as was seen in the last section, or they can come to act as upward channels for local influence on central government, as well as downward channels for central influence on local authorities. In this section, these two latter possibilities will be considered.

The central departments have to have access to those professional competences required in formulating and implementing government policy for locally delivered services. Unless they have access to these competences, especially in the more technical policy fields, the central policy makers can be at a serious disadvantage in relation to local authorities. The central departments do have some access to these competences through their contact with the officer advisers to the local authority associations; though consultation with these associations often has as much to do with defusing opposition as acquiring information. However, to continue to rely on this source of competences can place a central department in a position of at least potential dependence relative to local government. Even though these consultations were characterised by a comparatively high degree of trust, central departments were cautious about raising sensitive matters and mooting policy changes because they recognised that the prime loyalties of the officers involved were to their employers. Moreover, under the Conservative Government, with its radical right wing policies, this trust has lessened.

Central departments have sought, by recruiting their own professional staff, to reduce their dependence and the risks inherent in central-local consultations. Even though these staff may originate in local government, once they become departmental employees, a central department can expect them to owe their prime allegiance to the department as their employer. There seems to be a typical pattern of central government encouraging the development of a profession in local authorities and, once that profession has begun to establish itself, recruiting members of it; this pattern is apparent in the case of both professions studied.

Central departments require access to professional competences to perform three main functions: advice, regulation and promotion. Obviously, obtaining professional advice is particularly crucial in the early stages of policy making, as premature consultations with those in local government may create false expectations or enable opposition to be organised. For these reasons, the Housing Services Advisory Unit, particularly its Director, were drawn into policy discussions at ministerial level during the drafting of the Housing Act, 1980. Once policy decisions are made, regulation and promotion are important in getting new central initiatives off the ground. The difference between the two is, roughly, that between the stick and the carrot. Regulation involves placing constraints on authorities. In the case of roads, regulation was seen as a matter of applying technical competences, so the Department of Transport employed engineers to regulate the activities of local authorities in this field. In contrast, housing, referring here to the control of expenditure through the housing investment programmes, was regulated through generalist civil servants; though it should not be forgotten that the Department of the Environment also employed architects and surveyors at regional level to supervise some local authority housing design work.

Promotional functions mean presenting opportunities to local authorities, in the form of increased resources and in terms of new local policies or practices that they could adopt. One way of using increased resources in policy promotion is to offer special subsidies, for example the subsidy paid to authorities to encourage them to use high rise building systems during the late nineteen-sixties and early nineteen-seventies. However, more of the promotional policies in the housing issues examined in this study, did not involve using increased resources as a direct policy instrument. During the last Labour Government, local authorities were encouraged to play an interventionist role in the local housing system as they received increased powers relating to the private housing sector, as well as exhortations to intervene, including pushing housing investment programmes and the comprehensive housing service idea. Similarly, in housing management, the Department of the Environment took a much greater interest in a field traditionally seen as a local concern and promoted improvements in housing management practice such as through the demonstration Priority Estates Project.

In the roads sector, there are comparable instances of promotion by the Department of Transport. There was, through the introduction of the transport policies and programmes system and, later, the public transport plans, a central attempt to widen out the counties' conception of their role in local transport. More specifically, there has been a long term move towards the systematisation and standardisation of road maintenance practices and expenditure. The Marshall Report initiated this move, which was maintained by the DTp's continuing support through the Transport and Road Research Laboratory, the Standing Committee on Highway Maintenance and also the Department's insistence on county use of a standardised road assessment system in the maintenance of those trunk roads which they undertook under agency from the Department.

The professions can play a significant role in policy promotion. In fact, the initial recruitment of the members of a local government profession is almost invariably linked with major new policy initiatives directed towards local authorities; though such initiatives are not invariably associated with professional recruitment. In the 1920s one of the first actions of the government, when it took over national responsibility for major roads, was to recruit some senior local government engineers. Similarly, during the middle nineteen-seventies, the DoE's increased interest in local housing policies was reflected in the recruitment of three senior housing officers as housing advisers.

In housing, the profession played a limited promotional role on behalf of central government. The professional community did push the idea of a comprehensive housing service, for obvious reasons of professional self-interest, though a significant minority were unhappy with the CHS idea. But local policy initiatives, in line with the progressive approach favoured by the Department, tended to be the product of local circumstances - sometimes the elected members took the lead and other times the chief housing officer.

The highway engineers were much more important than the housing profession in supporting and promoting policy change in line with central government intentions. For professional reasons, it should be said, and not out of any deference towards the DTp. On road maintenance, in particular, the professionals, mainly the younger professionals, enthusiastically supported the initial introduction of standards and road assessment systems. But it should be noted that the interest in maintenance was heightened when it became clear that the funds available for road construction were falling.

Within the central-local government relationship, then, central departments employ members of the local government professions to control and influence local authorities. Typically, professionals strongly support the presence of fellow professionals within central departments. Both the road engineers and the housing professionals supported the appointment of fellow professionals and this would seem to be true in the case of other professions, like social work and education. Professionals see the presence of professional colleagues in central government as a recognition of professional status and those professionals as, at least potential, allies. It should be said, too, that the local authority associations have generally favoured the recruitment of professionals by central government not for the same reasons but because it has seemed to them likely to improve central government understanding of local problems. These professional links can assist joint working, particularly as long as the policy aims of local authorities and central government are more or less congruent. However, where policy aims diverge, the availability of professional resources within a central department enables it to dominate local policies more effectively than it might otherwise do. Moreover, the employment relationship tends to be given precedence by professionals over the collegial relationship. Those professionals employed by the centre are generally more likely to identify their interests with their department than with the local authorities. Thus, for example, the civil service engineers favoured the increased privatisation of trunk road construction because it appeared to them to strengthen their control over projects; meanwhile, the engineers in the counties strongly opposed this development, which they recognised as reducing their own influence over road construction projects. In fact, as was seen in the case of highway agency agreements, the employment relationship and organisational loyalties are a major source of conflict within the engineering profession.

THE PROFESSIONS AND CHANGE

The discussion so far has raised a number of questions about the relationship of the professions to policy change and innovation. Professions have been seen to be both agents of change on behalf of central government and in their own right, and also resistant to change when it is perceived as threatening their position and, indeed, that of their employers. There is a paradox here. Although, in general policy terms, the professions, once established, tend to be forces for conservatism and policy maintenance, in narrower terms they can also act as forces for change. These latter changes are typically intended to make existing policies more workable and to absorb new problems into the professional account. Thus they can be thought of as instances of what Schon calls 'dynamic conservatism'.[8] In this section what this means for policy will be considered.

Firstly, professions can act as forces resistant to change in a number of ways. Quite simply, their existence in government organisations necessarily imposes constraints on the deployment of manpower and so on the pace of policy change. Housing professionals cannot easily be moved to supervise trunk road maintenance or highway engineers to assess local authorities' housing plans. The professionals themselves usually stress these barriers to manpower movement, emphasising the high level of cognitive competences required for given tasks. The arguments are difficult for lay people to challenge. This problem of the non-interchangeable nature of professional competences is further exacerbated by professions' organised opposition to changes likely to reduce their role and influence in policy making and implementation. The ability of professions to use the organisational resources available to them is well illustrated by the road engineers' counter attack on the growing environmental lobby.

Secondly, the professions are also associated with change and innovation. It has been frequently argued that, as a form of social organisation, the profession provides incentives for practitioners to search for and experiment with new ideas, to exchange such ideas and to support innovation on behalf of colleagues, and a collegial network through which new ideas are diffused. The incentives for practitioners to engage in these activities are enhanced professional reputations and the increased material and status rewards following from that, particular career promotion, as well as the fulfillment of a presumed moral obligation to the employer or client. This innovating role can be divided into three further roles - the originating, the diffusing and the supporting.

The purely originating role is certainly ascribed considerable importance in the rhetoric of professions. Most professional associations place an emphasis on research by having research committees, maintaining contacts with universities and other research establishments, and even carrying out their own research. However, the research functions of the professional associations tend to be more related to diffusion than to origination. Most innovations originate outside these associations. Major sources of innovation are the professionals working in local authorities. Thus, the original idea of, and experiments with, road assessment systems came from a county council and were then taken up, developed and diffused by the Transport and Road Research Laboratory (TRRL) and the County Surveyors Society. Another source is an organisation like the TRRL which can claim to have originated a number of innovations in the roads field. Yet another source is the private sector. Indeed, in the highways field there are a number of examples of a three-way relationship among the TRRL, private firms and local authorities in the development of new materials and equipment. All three have, of necessity, to work together. For example, the development of a new road surface is an interactive process involving all three - local authorities define the broad requirements and provide sites for testing and the promise of future markets, the TRRL the scientific advice, and private companies the development funds and investment in prototype materials and machines. Under these circumstances, the professional engineers are in a mediating position between their authorities and private firms. It has been suggested that where professionals occupy such a mediating role, the private sector comes to exercise undue influence over the sector; in other words, the private sector defines the requirements of the public organisations.[9] The evidence in this study neither supports nor refutes this suggestion. It is possible that there is an over-investment in the search for new materials and equipment. However, if anything, the indications are that there has been under-investment. The problem with major roads, particularly with motorways, has been that the road surfaces have been short lived, especially on the early motorways. At least, in part, this was because the government of the time went for the cheapest design specifications based on conservative figures of future road use. In consequence, these roads have had to be resurfaced much sooner than was originally anticipated and the motorways have continued to be dogged by surfacing problems.

There is another non-professional source of innovation that should be mentioned - the elected members themselves. Obviously, in a highly technical field like highway engineering, there are obstacles to the contributions they may make to innovation; though they may have strong views on the route of a new road, they are unlikely to have strong views, or any views, on type of surface to be used. The elected members see little point in incurring the costs in time and effort involved in trying to overcome these technical barriers. However, the situation is very different in the case of housing, where these costs are considerably lower. Moreover, housing is usually seen as politically very significant and politicians are typically closely involved in housing policy. Indeed, many of the major innovations in housing have come from the politicians. Thus most of the experiments and initiatives in tenants' participation and rights were created by the elected members; Basildon is a good example, where the Leader of the Council initiated their tenants' charter scheme. It should be added that the interest of local and national Labour Party politicians in the tenants' charter reflected, in part at least, a concern that they had to have an electorally appealing answer to the housing sales policy of the Conservatives. Conservative politicians, too, have contributed to the pool of new ideas in housing, homesteading being one example.

The diffusion of innovations is as important as actual origination. Professions can form potent mechanisms for diffusion as they provide incentives and collegial networks along which information can be exchanged. Engineering at the county level is an excellent example of how professional organisations can work as diffusionary mechanisms, as was seen in the chapter on highway maintenance, there was an intricate network of professional associations and working parties. This collegial network was strengthened by career movement from central into local government as the roads programme expanded and as the road construction units came into existence as horizontal organisations bridging local and central government.

In housing, the professional collegial network was very attenuated, if not non-existent. A collegial network, comparable to that enjoyed by the road engineers, had not developed within the profession. The essential precondition for the development of such a network, professional cohesiveness, had failed to develop. So the factors which had inhibited the development of such cohesiveness had effectively inhibited the growth of a collegial network.

These collegial networks are also significant in providing assurances of mutual support to officers who might otherwise feel isolated and threatened within their employing authorities. This sense of threat or anxiety seems likely to have played a part in encouraging the formation of professional groupings within local government, as officers made contact with others doing similar tasks in other authorities. The mutual support that such contacts can give, offers the psychological strength for officers to push for change in their own authority. These networks may also tend to contribute towards the tendency for local policies to converge.

CONCLUSION

This study has attested to both the weaknesses and strengths of professions within the context of local and central government. The aim has been to avoid rhetorical statements about professional power but rather to examine such power through detailed investigations of the professionalisation process and the policy influence of professions. The study has indicated that professions can, and do, have a significant influence over policy, so that there is some basis for the often expressed concerns over the professional arrogation of policy making. Professions, once they are established, do create and sustain their own policy dynamic. But this dynamic is by no means unstoppable. Major changes in the political and economic climates affect the standing of a profession in the view of others and so reduce their influence. Thus, even the well-organised road engineers had to accommodate themselves to the changing availability of resources for road construction and maintenance and to changes in policy views and scepticism towards professionals.

The housing profession has made an instructive contrast to the highway engineers. The limited progress of the profession illustrates some of the obstacles to professionalisation and to professional influence over policy. The housing policy sector was one which the elected politicians and, to a lesser extent, central government civil servants dominated and the influence of the professions was consequently very constrained. The arrested attempts of housing to become further professionalised points to a change of attitude towards professionalisation among local and central policy makers. This and other political developments suggest that what had come to seem like a continuing trend toward professionalisation across all policy sectors has been halted and, possibly, even reversed.

NOTES

[1] It should be noted that the engineering profession or, for that matter, the civil engineering profession, as a whole has not exhibited the same degree of group cohesion but has been fragmented into specialisms.

[2] Eliot Freidson, The Profession of Medicine, Dodd Mead, New York, 1970 and other works cited in chapter three; and Magali Sarfati Larson, The Rise of Professionalism, University of California Press, Berkeley, 1977.

[3] See G E Cherry, The Evolution of British Town Planning, Leonard Hill, London, 1974, pp.165-9.

[4] See Phoebe Hall's account of these events in Reforming the Welfare : The Politics of Change in the Personal Social Services, Heinemann, London, 1976.

[5] For a recent survey of the road lobby see G Dudley, 'The Road Lobby: A Declining Force?' in David Marsh (ed), Pressure Politics : Interest Groups in Britain, Junction Books, London, 1983, pp.104-127.

[6] Martin Painter, 'Whitehall and Roads: A Case Study of Sectorial Politics', Policy and Politics, 8,2, April 1980, p.172.

[7] County surveyor in personal interview.

[8] Donald A Schon, Beyond the Stable State, Temple Smith, London, 1971, Ch.2.

[9] For instance, Patrick Dunleavy, Urban Political Analysis, MacMillan, London, 1980, pp.113-7.